Modern Raman Spectroscopy

Modern Raman Spectroscopy: A Practical Approach

SECOND EDITION

Ewen Smith
Emeritus Professor, University of Strathclyde, UK

Geoffrey Dent
GD Analytical Consulting, and University of Manchester, UK

Registered Offices
John Wiley & Sons, Inc., 111 River Street, Hoboken, NJ 07030, USA
John Wiley & Sons Ltd, The Atrium, Southern Gate, Chichester, West Sussex, PO19 8SQ, UK

Editorial Office
The Atrium, Southern Gate, Chichester, West Sussex, PO19 8SQ, UK

For details of our global editorial offices, customer services, and more information about Wiley products visit us at www.wiley.com.

Wiley also publishes its books in a variety of electronic formats and by print-on-demand. Some content that appears in standard print versions of this book may not be available in other formats.

Library of Congress Cataloging-in-Publication Data

Names: Smith, Ewen, author. | Dent, Geoffrey, 1948– author.
Title: Modern Raman spectroscopy : a practical approach / Ewen Smith,
  Geoffrey Dent.
Description: Second edition. | Hoboken, NJ : Wiley, [2019] | Includes
  bibliographical references and index. |
Identifiers: LCCN 2018042658 (print) | LCCN 2018051812 (ebook) | ISBN
  9781119440581 (Adobe PDF) | ISBN 9781119440543 (ePub) | ISBN 9781119440550
  (hardcover)
Subjects: LCSH: Raman spectroscopy.
Classification: LCC QD96.R34 (ebook) | LCC QD96.R34 S58 2019 (print) | DDC
  535.8/46–dc23
LC record available at https://lccn.loc.gov/2018042658

Cover Design: Wiley
Cover Image: © Veronaa/Getty Images

Set in 10/12pts Times by SPi Global, Pondicherry, India

10 9 8 7 6 5 4 3 2 1

Contents

Preface

Since the first edition of our book, there has been a huge expansion in the use of Raman spectroscopy. Advances in optics, electronics and data handling combined with improvements made by manufacturers and spectroscopists have made Raman scattering easier to record and more informative. Small, portable spectrometers are rugged, reliable and are becoming less expensive. Some can work powered by low-voltage (1.5 V) batteries and give good performance in hostile environments. At the other end of the scale, advanced equipment is simpler, more sensitive, more flexible and more reliable. New methods with improved performance have been developed. As a result, a technique which was once labelled by some as lacking sensitivity can now, in the correct form, probe the electronic structure of a single molecule or be used to help in the diagnosis of cancer. This has attracted many more users into the field with a wide range of backgrounds.

Our aim in writing this book is to provide the understanding necessary to enable new users to apply the technique effectively. In the early chapters we provide basic theory and practical advice to enable the measurement and interpretation of Raman spectra with the minimum barrier to getting started. However, for those with a deeper understanding of the effect, Raman scattering is a very rich technique capable of providing unique information and a unique insight into specific problems. In writing this book some difficult choices have had to be made around the presentation of the theory, particularly with the wide variety of backgrounds we expect readers to possess. We have used as few equations as possible to show how the theory is developed and those are deliberately placed after the chapters on basic understanding. We concentrate on molecular polarizability, the molecular property which controls intensity. The equations are explained, not derived, so that those with little knowledge of mathematics can understand the conclusions reached and those of a more mathematical bent can use the framework for further investigation. This enables selection rules, resonance Raman scattering and some of the language in modern literature to be understood. This is not the traditional approach but, although deriving scattering theory from first principles is good for understanding, it adds little to Raman interpretation. Classical theory which does not use quantum mechanics cannot deliver the information required by most Raman spectroscopists. For these reasons, references to these areas are given but the theory is not explained.

Surface-enhanced Raman scattering accounts for a significant fraction of the papers on Raman scattering and is employed in quite a few modern developments, so is given a full chapter. We finish with two chapters designed to enable the reader to come up to speed on the way Raman scattering is now applied in some of the main fields and to introduce the new techniques that are providing key insights and greater performance. Some of the new techniques are still expensive and therefore not so widely available but if the improvements continue, as is likely, they will become much more accessible. In any case the reader should be aware of their advantages.

One of the practical difficulties faced is in compliance with the IUPAC convention in the description of spectrum scales. The direction of the wavenumber shift should be consistent but this is not always the case in the literature. Further, Raman scattering is a shift from an exciting frequency and should be labelled Δcm^{-1} but it is common practice to use cm^{-1} with the delta implied. As far as possible we have used the format in which the user is most likely to record a spectrum or to see it in the literature. However, where we have used literature examples in this book, it is not possible to change these. We apologize to the purists who would prefer complete compliance with the IUPAC convention, but we have found that this is not practicable.

The authors hope that those who are just developing or reviving an interest in Raman spectroscopy will very quickly gain a practical understanding from the first two chapters. Furthermore, it is hoped that they will be inspired by the elegance and information content of the technique to delve further into the rest of the book, and explore the vast potential of the more sophisticated applications of Raman spectroscopy.

Acknowledgements

We thank professors Duncan Graham and Karen Faulds and members of the Centre for Molecular Nanometrology at Strathclyde University and some members of the older Raman group for supplying diagrams, the proprietors of the Analytical Sciences functions in Blackley, Manchester, UK for their use of facilities and permissions to publish material generated and our respective wives, Frances and Thelma, for putting up with us.

Chapter 1

Introduction, Basic Theory and Principles

1.1 INTRODUCTION

The main spectroscopies employed to detect vibrations in molecules are based on the processes of infrared absorption and Raman scattering. They are widely used to provide information on chemical structures and physical forms, to identify substances from the characteristic spectral patterns ('fingerprinting') and to determine quantitatively or semiquantitatively the amount of a substance in a sample. Samples can be examined in a whole range of physical states, for example, as solids, liquids, vapours, hot and cold, in bulk, as microscopic particles or as surface layers. The techniques are very wide ranging and provide solutions to a host of interesting and challenging analytical problems. Raman scattering is less widely used than infrared absorption, largely due to problems with sample degradation and fluorescence. However, recent advances in instrument technology have simplified the equipment and reduced the problems substantially. These advances, together with the ability of Raman spectroscopy to examine aqueous solutions, samples inside glass containers and samples without any preparation, have led to a rapid growth in the application of the technique.

Practically, modern Raman spectrometers are simple to use. Variable instrument parameters are few, spectral manipulation is minimal and a simple interpretation of the data may be sufficient. This chapter and Chapter 2 aim to set out the basic principles and experimental methods to give the reader a firm understanding of the basic theory and practical considerations so that the technique can be applied at the level often required for current applications. However, with Raman scattering important information is sometimes not used or recognised. Later chapters will develop the minimum theory required to give a more in-depth understanding of the data obtained

Modern Raman Spectroscopy: A Practical Approach, Second Edition. Ewen Smith and Geoffrey Dent.
© 2019 John Wiley & Sons Ltd. Published 2019 by John Wiley & Sons Ltd.

and to enable comprehension of some of the many more advanced techniques, which have specific advantages for some applications.

1.2 HISTORY

The phenomenon of inelastic scattering of light was first postulated by A. Smekal in 1923 [1] and first observed experimentally in 1928 by C.V. Raman and K.S. Krishnan [2]. Since then, the phenomenon has been referred to as Raman scattering. In the original experiment, sunlight was focused by a telescope onto a sample, which was either a purified liquid or a dust-free vapour. A second lens was placed by the sample to collect the scattered radiation. A system of optical filters was used to show the existence of scattered radiation with an altered frequency from the incident light – the basic characteristic of Raman scattering.

1.3 BASIC THEORY

When light interacts with matter, the photons which make up the light may be absorbed or scattered or may not interact with the material and may pass straight through it. If the energy of an incident photon corresponds to the energy gap between the ground state of a molecule and an excited state, the photon may be absorbed and the molecule promoted to the higher energy excited state. It is this change which is measured in absorption spectroscopy by detection of the loss of that energy of radiation. However, it is also possible for the photon to interact with the molecule and scatter from it. In this case there is no need for the photon to have an energy which matches the difference between two energy levels of the molecule. The scattered photons can be observed by collecting them at an angle to the incident light beam. If there is no absorption from any electronic transition, which has a similar energy to that of the incident light, the scattering efficiency increases as the fourth power of the frequency of the incident light.

Scattering is a commonly used technique. For example, it is widely used as a method for measuring particle size and size distribution down to sizes less than 1 μm. One everyday illustration of this is that the sky is blue because the higher energy blue light is scattered from molecules and particles in the atmosphere more efficiently than the lower energy red light. However, for molecular identification, a small component of the scattered light, Raman scattering, is particularly effective.

The word 'light' implies electromagnetic radiation within the wavelength range to which the eye is sensitive whereas in spectroscopy, the important point is whether the detector is sensitive to the radiation used and as a consequence much wider ranges of wavelengths can be used. As a result, the process of absorption is used in a wide range of spectroscopic techniques. For example, it is used in acoustic spectroscopy where

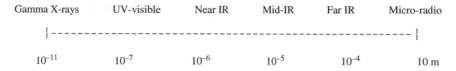

Figure 1.1.   The electromagnetic spectrum on the wavelength scale.

there is a very small energy difference between the ground and excited states and in X-ray absorption spectroscopy where there is a very large difference. In between these extremes, many of the common techniques such as NMR, EPR, infrared absorption, electronic absorption and fluorescence emission and vacuum ultraviolet spectroscopy are based on the absorption of radiation. Figure 1.1 indicates the wavelength range of some commonly used types of radiation.

Radiation is often characterised by its wavelength (λ). However, in Raman spectroscopy, we are interested in information obtained from the scattered radiation on the vibrational states of the molecule being examined. These are usually more conveniently discussed in terms of energy and consequently it is usual to use frequency (ν) or wavenumber (ϖ) scales, which are linearly related to energy. The relationships between these scales are given in the equations below.

$$\lambda = \frac{c}{\nu} \tag{1.1}$$

$$\nu = \frac{\Delta E}{h} \tag{1.2}$$

and

$$\varpi = \frac{\nu}{c} = \frac{1}{\lambda} \tag{1.3}$$

It is clear from Eq. (1.1) that the energy is proportional to the reciprocal of wavelength and therefore the highest energy region is shown on the left side in Figure 1.1.

The energy changes we detect in vibrational spectroscopy are those required to cause nuclear motion but the way in which radiation is employed in infrared and Raman spectroscopies is different. In infrared spectroscopy, infrared energy covering a range of frequencies is directed onto the sample. Absorption occurs where the frequency of the incident radiation matches that of a vibration so that the molecule is promoted to a vibrational excited state. The loss of this frequency of radiation from the beam after it passes through the sample is detected. In contrast, Raman spectroscopy uses a single frequency of radiation to irradiate the sample. It is the radiation scattered from the molecule, one vibrational unit of energy different from the incident beam, which is detected. Thus, unlike infrared adsorption,

Raman scattering does not require matching of the incident radiation to the energy difference between the ground and excited states.

In a scattering process, the light interacts with the molecule and distorts (polarizes) the cloud of electrons round the nuclei to form a short-lived state called a virtual state discussed in Chapter 3. This state is not stable and the photon is quickly reradiated. If only the electron cloud is distorted in the scattering process, when the electron cloud returns to the starting position the photons are scattered with the same frequency as the incident radiation. This scattering process is regarded as elastic scattering and is the dominant process. For molecules, it is called Rayleigh scattering. However, if nuclear motion is induced during the scattering process, energy will be transferred either from the incident photon to the molecule or from the molecule to the scattered photon. In these cases, the process is inelastic and the energy of the scattered photon is different from the incident photon by one vibrational unit. This is Raman scattering. It is inherently a weak process in that only one in every 10^6–10^8 photons which scatter is Raman scattered. In itself this does not make the process insensitive since with modern lasers and microscopes very high-power densities can be delivered to very small samples but it does follow that other possible processes such as sample degradation and fluorescence can occur readily.

Figure 1.2 shows the basic processes which occur for one vibration. At room temperature, most molecules, but not all, are present in the lowest energy

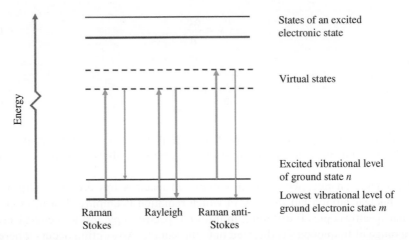

Figure 1.2. Diagram of the Rayleigh and Raman scattering processes. The lowest energy vibrational state m is shown at the foot with a state one vibrational unit in energy above it labelled *n*. Both the excitation energy (upward arrows) and the scattered energy (downward arrows) have much larger energies than the energy of a vibration. Two levels from the next highest electronic state are shown. Clearly with radiation of the frequency used, absorption of the exciting radiation would not occur. Rayleigh scattering also occurs from higher vibrational levels such as *n*.

vibrational level. The virtual states are not real states of the molecule but are created when the laser interacts with the electrons and causes polarization and the energy of these states is determined by the frequency of the light source used. The Rayleigh process will be the most intense process since most photons scatter this way. Since it does not involve any energy change, the diagram shows the light returning to the same energy state. The Raman scattering process from the ground vibrational state m leads to absorption of energy by the molecule and its promotion to the higher energy excited vibrational state n. This is called Stokes scattering. However, due to thermal energy, some molecules may be present initially in an excited state as represented by n in Figure 1.2. Scattering from these states to the ground state m is called anti-Stokes scattering and involves transfer of energy from the molecule to the scattered photon. The relative intensities of the two processes depend on the population of the various states of the molecule and also on symmetry selection rules. The populations can be worked out from the Boltzmann equation (see Chapter 3) but at room temperature, the number of molecules expected to be in an excited vibrational state other than really low-energy states will be small.

Thus, compared to Stokes scattering, anti-Stokes scattering will be weak and, in general, will become weaker the higher the energy of the vibration, due to the decreasing population of the excited vibrational states. Further, anti-Stokes scattering will increase relative to Stokes scattering as the temperature rises [3]. Figure 1.3 shows a typical spectrum of Stokes and anti-Stokes scattering from carbon tetrachloride separated by the intense Rayleigh scattering, which is off scale close to the point where there is no energy shift. Note that no signal is shown in the spectrum in the low-frequency region because a filter has been added in front of the spectrometer to

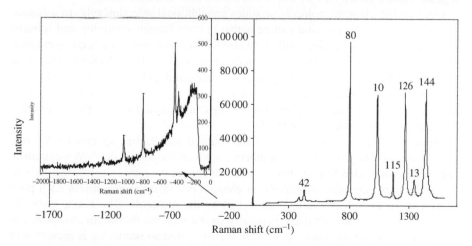

Figure 1.3. Stokes and anti-Stokes scattering for cyclohexane. To show the weak anti-Stokes spectrum, the *y*-axis has been extended in the inset.

remove almost all light within about $200\,cm^{-1}$ of the exciting line. Some breakthrough of the laser light can be seen where there is no energy shift at all.

Usually, Raman scattering is recorded only on the low-energy side to give Stokes scattering but on occasion anti-Stokes scattering is preferred. For example, where there is fluorescence interference this will occur at a lower energy than the excitation frequency and consequently anti-Stokes scattering can be used to avoid interference. The difference in intensities of the bands in Stokes and anti-Stokes scattering for any one vibration can also be used to measure the temperature.

Figure 1.2 illustrates one key difference between infrared absorption and Raman scattering. As described above, infrared absorption would involve direct excitation of the molecule from state m to state n by a photon of exactly the energy difference between them. In contrast, Raman scattering uses much higher energy radiation and measures the difference in energy between n and m by subtracting the scattered photon energy from that of the incident beam (the two vertical arrows in each case).

The cyclohexane spectrum in Figure 1.3 shows that there is more than one vibration which gives effective Raman scattering (i.e. is Raman active) and the nature of these vibrations will be discussed in the next section. However, there is a basic selection rule which is required to understand this pattern. Intense Raman scattering occurs from vibrations which cause a change in the polarizability of the electron cloud round the molecule. Usually symmetric vibrations cause the largest changes. This contrasts with infrared absorption where the most intense absorption is caused by a change in dipole and hence asymmetric vibrations which cause this are the most intense. As will be seen later not all vibrations of a molecule need, or in some cases can, be both infrared and Raman active and the two techniques usually give quite different intensity patterns. As a result, the two are often complementary and, used together, give a better view of the vibrational structure of a molecule.

One specific class of molecule provides an additional selection rule. In a centrosymmetric molecule, no band can be active in both Raman scattering and infrared absorption. This is sometimes called the *mutual exclusion rule*. In a centrosymmetric molecule, reflection of any point through the centre will reach an identical point on the other side (planar C_2H_4 is centrosymmetric, tetrahedral CH_4 is not). This distinction is useful particularly for small molecules where a comparison of the spectra obtained from infrared absorption and Raman scattering can be used, for example, to differentiate *cis* and *trans* forms of a molecule.

Figure 1.4 shows a comparison of the infrared and Raman spectra for benzoic acid. The x axis is given in wavenumbers for which the unit is cm^{-1}. Wavenumbers are not recommended SI units but the practice of spectroscopy is universally carried out using these and this is unlikely to change. For infrared absorption each peak represents an energy of radiation absorbed by the molecule. The y-axis gives the amount of the light absorbed and is usually shown with the maximum absorbance as the lowest point on the trace. As is often the case, Raman scattering is presented in Figure 1.4 only as the Stokes spectrum and along the x-axis each vibration is presented as the shift in energy from the energy of the laser beam. In this way the difference

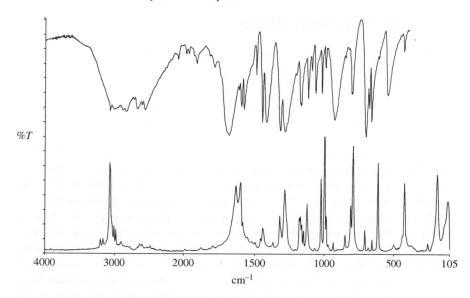

%T

4000   3000   2000   1500   1000   500   105

cm⁻¹

Figure 1.4. Infrared and Raman spectra of benzoic acid. The top trace is infrared absorption given in % transmission (%*T*) so that the lower the transmission value the greater the absorption. The lower trace is Raman scattering and the higher the peak the greater the scattering.

in energy between the ground and excited vibrational states for each vibration (*n* and *m* in Figure 1.2) is shown in the spectrum.

Strictly speaking, since Raman scattering is expressed as a shift in energy from that of the exciting radiation, it should be referred to as Δcm^{-1} but it is often expressed simply as cm^{-1}. This practice is followed in this book for simplicity. The information of interest, to most users, is in the 3600–400 cm^{-1} (2.8–12 μm) range commonly used in infrared spectroscopy since this includes most modes, which are characteristic of a molecule. In some applications, much larger or smaller energy changes are studied and modern Raman equipment can cope with much wider ranges. One specific advantage of Raman scattering is that shifts down to about 100–200 cm^{-1} can easily be recorded and, with the correct equipment, much smaller shifts can be measured so that features such as lattice vibrations can be studied.

The intensities of the bands in the Raman spectrum are dependent on the nature of the vibration being discussed and on instrumentation and sampling factors. Modern instruments should be calibrated to remove the instrument factors but this is not always the case and these factors are dealt with in the next chapter. Sampling can have a large effect on the absolute intensities, band widths observed and band positions. Again these will be dealt with later. This chapter will set out a step-by-step approach to interpreting the Raman scattering from the set of vibrations present in a molecule without reference to instrumental or sampling factors.

1.4   MOLECULAR VIBRATIONS

If there is no change in electronic energy, for example, by absorption of a photon and the promotion of an electron to an excited electronic state, the energy of a molecule can be divided into a number of different parts or 'degrees of freedom'. Three of these degrees of freedom are taken up to describe the translation of the molecule in space and three to describe rotational movement except for linear molecules where only two types of rotations are possible. The rest are vibrational degrees of freedom or in other words the total number of vibrations which could appear in a spectrum. Thus, if N is the number of atoms in a molecule, the number of vibrational degrees of freedom and therefore the number of vibrations possible is $3N - 6$ for all molecules except linear ones where it is $3N - 5$. For a diatomic molecule, this means there will be only one vibration. In a molecule such as oxygen, this is a simple stretch of the O—O bond. This will change the polarisability of the molecule but not the dipole moment since there is no dipole in the molecule and the vibration is symmetric about the centre. Thus, the selection rules already discussed would predict, and it is true, that oxygen gas will give a band in the Raman spectrum and no bands in the infrared spectrum. However, in a molecule such as nitric oxide, NO, there will be only one band but, since there is both a dipole change and a polarizability change, it will appear in both the infrared and Raman spectrum.

A triatomic molecule will have three modes of vibration. They are a symmetrical stretch, a bending or deformation mode and an anti-symmetrical (often referred to as asymmetrical stretch, as shown). Figure 1.5 shows these for water (H_2O) and carbon dioxide (CO_2).

These diagrams use 'spring and ball' models. The spring represents the bond or bonds between the atoms. The stronger the bond the higher the frequency. The balls represent the atoms and the heavier they are the lower the frequency. The expression

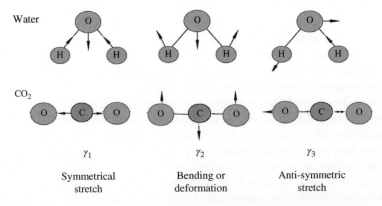

Figure 1.5.   Spring and ball models for three modes of vibration for H_2O and CO_2.

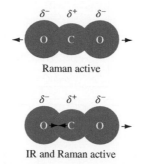

Figure 1.6. Electron cloud model of water and carbon dioxide.

which relates the mass of the atoms and the bond strength to the vibrational frequency is Hooke's law, which is dealt with in Chapter 3, but for the present, it is clear that strong bonds and light atoms will give higher frequencies of vibration and heavy atoms and weak bonds will give lower ones.

This simple approach of estimating the likely frequencies of a vibration based on the atoms and the nature of the bonds is widely used to interpret vibrational spectra and is backed up by an extensive literature which has enabled the frequency range in which common vibrations are likely to appear has been defined. However, the intensities of the bands are also important. The molecule actually exists as a three-dimensional structure with a pattern of varying electron density covering the whole molecule. A simple depiction of this for the carbon dioxide and water molecules is shown in Figure 1.6. When the molecule is vibrating, the electron cloud will alter as the positive nuclei change position and depending on the nature of the change this can cause a change of dipole moment or polarization.

In these triatomic molecules, the symmetrical stretch causes large polarization changes and hence strong Raman scattering with weak or no dipole change and hence weak or no IR absorption. The deformation mode causes a dipole change but little polarisation change and hence strong IR absorption and weak or nonexistent Raman scattering. However, it should be noted that from the $3N - 5$ rule there should be four vibrations. There are two bending modes, only one of which has been illustrated. Figure 1.7 illustrates the vibrations possible for carbon disulphide along with the corresponding IR absorption and Raman scattering spectra.

Although this type of analysis is suitable for small molecules, it is more difficult to apply in a more complex molecule. Figure 1.8 shows one vibration from a dye in which a large number of atoms are involved. This is obtained from a theoretical calculation using density functional theory (DFT), which is discussed briefly later. It probably gives a depiction of the vibration, which is close to the truth. However, even if it were possible to calculate the spectrum of every molecule quickly in the laboratory, which at present it is not, this type of diagram is only of limited utility to the spectroscopist. A comparison between molecules of similar type is difficult

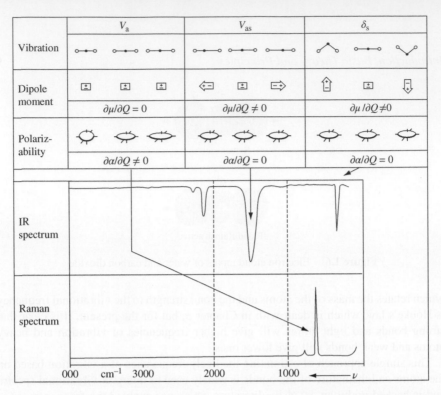

Figure 1.7. Dipole and polarisation changes in carbon disulphide, with resultant infrared and Raman spectra. Source: Reprinted from Fadini, A. and Schnepel, F.-M. (1989). *Vibrational Spectroscopy: Methods and Applications*. Chichester: Ellis Horwood Ltd [4].

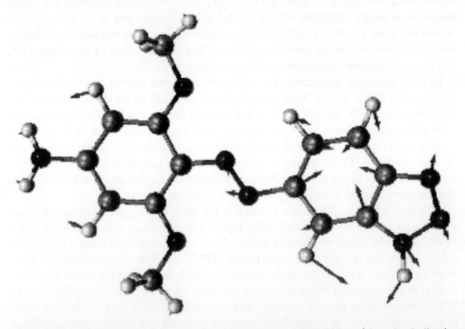

Figure 1.8. A displacement diagram for a vibration at about $1200\,cm^{-1}$ in a dye indicating the involvement of a number of atoms. The arrows show the direction of the displacement. Since the equilibrium position of the atoms is shown, during a complete vibration the arrows will reverse in direction.

unless a full calculation is available for all of them and each subtle change in the nuclear displacements is drawn out or accurately described for each one. This limits the ability to compare large numbers of molecules or to understand the nature of vibrations in molecules for which there is no calculation.

The usual approach to describing vibrations is to simplify the problem and break the displacements down into a number of characteristic features, which can relate to more than one molecule. In the vibration shown in Figure 1.8, the biggest displacements of the heavier atoms is on one of the ring systems and the vibration would almost certainly be labelled vaguely as a 'ring stretch'. In another vibration, not shown, the situation was much simpler. Large displacements were found on the two nitrogen atoms which form the azo bond between the rings and the direction indicated bond lengthening and contracting during the vibrational cycle. Thus, this vibration is called the azo stretch. There is a change in polarisability just as there was for oxygen, so it should be a Raman-active vibration. We can search for these vibrations in the actual spectrum and hopefully match a peak to the vibration. This is called assigning the vibration. Thus, it is possible to describe a vibration in a few helpful words. In some cases this is fairly accurate as for the azo stretch but in some cases, the description is not adequate to describe the actual movement as 'ring stretch' would be for the vibration shown. However, common bands can be assigned reasonably by this method for many molecules making communication easier but it is best to be aware that this only a rough approximation in some cases.

1.5 GROUP VIBRATIONS

Before assigning vibrations to peaks in the spectrum it is necessary to realise that two or more bonds which are close together in a molecule and are of similar energies can interact and it is the vibration of the group of atoms linked by these bonds that is observed in the spectrum. For example, the CH_2 group is said to have a symmetric and an anti-symmetric stretch rather than two separate CH stretches (Figure 1.9). It follows from this that different types of vibrations are possible for different groups and these require to be identified in the spectrum. Selected examples of a few of these for $-CH_3$ groups and benzene are shown in Figure 1.9 to illustrate this.

In contrast, where there is a large difference in energy between the vibrations in different bonds or if the atoms are well separated in the molecule, they can be treated separately. Thus, for CH_3Br, the C—H bonds in CH_3 must be treated as a group but the C—Br vibration is treated separately since it is at a much lower energy due to the heavy bromine atom (Figure 1.9). The selected vibrations of benzene are shown in two different ways. First, they are shown with the molecule in the equilibrium position with arrows showing the direction of the vibrational displacement. To illustrate what this means, they are also shown with the vibration at the extremes of the vibrational movement.

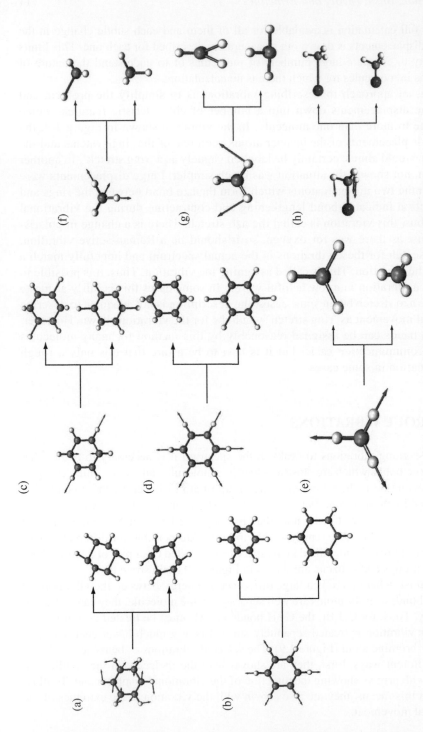

Figure 1.9. Selected displacement diagrams for benzene and for CH_3 in CH_3Br. (a) A quadrant stretch for benzene at about $1600\,cm^{-1}$. (b) The symmetric breathing mode at just above $1000\,cm^{-1}$. (c and d) Two C—H vibrations at about $3000\,cm^{-1}$. (e) The symmetric stretch of CH_3 in CH_3Br at above $2800\,cm^{-1}$. (f) An asymmetric stretch at above $2900\,cm^{-1}$. (g) A CH band at about $2900\,cm^{-1}$. (h) A low-frequency mode at below $600\,cm^{-1}$.

1.6 BASIC INTERPRETATION OF A SPECTRUM

It is possible to give energy ranges in which the characteristic frequencies of the most common groups which are strong in Raman scattering can occur. The relative intensities of specific peaks help to confirm that the correct vibration has been picked out. For example, carbonyl groups >C=O are usually present at ~1700 cm^{-1}. They have a dipole moment which will change when the group stretches and there will be some polarisation change. Thus, they give strong bands in the infrared spectrum and weaker bands in the Raman spectrum. Symmetrical groups such as unsaturated bonds (—C=C—) and disulphide bonds (—S—S—), which have a significant polarization change, are strong Raman scatterers and weak infrared absorbers. The stretching mode for these vibrations is ~1640 and 500 cm^{-1}, respectively. There are many more examples. It is the combination of knowledge of approximate frequency and likely relative intensity of particular vibrations which form the basis of the assignment process used by most spectroscopists. For example, the 4000–2500 cm^{-1} is the region where single bonds consisting of light elements (e.g. C—H and N—H) scatter. Aromatic C—H stretches are above 3000 cm^{-1} and aliphatic ones below it. The 2700–2000 cm^{-1} range is where vibrations of multiple bonds such as —N=C=O and triple bonds like ethynes and cyanide appear. The 2000–1500 cm^{-1} region is where double bonds (—C=O, —C=N, —C=C—) occur. The 1600–1000 cm^{-1} region is generally referred to as the fingerprint region. Some groups, e.g. nitro (O=N=O), do have specific bands in this region but mainly it is dominated by C—C bonds and associated C—H bonds, etc. Phenyl rings have vibrations between 1600 and 1550 cm^{-1} due to quadrant stretching modes and a totally symmetric mode at about 1000 cm^{-1} as shown in Figure 1.9. In addition, there are semicircular modes at about 1300 cm^{-1}. Bands below 650 cm^{-1} usually arise in specific groups like S—S or C—I vibrations, more complex often out-of-plane vibrations in larger molecules, inorganic groups, metal–organic groups or lattice vibrations. Tables 1.1–1.5 show the frequency ranges of many of the vibrations, which give rise to strong bands in Raman spectroscopy. The ranges are approximate for the groups in most structures but some groups in unusual structures may give bands outside these ranges. The position of the band across the chart is only for clarity of illustration and does not imply strength. The thickness of the line indicates the relative strength. These tables enable a beginning to be made on the assignment of specific bands.

A more difficult problem is in estimating the relative intensities of the bands. Earlier, we showed that there are reasons why in some circumstances bands which are strong in the infrared spectrum are not strong in the Raman spectrum and vice versa. Although this cannot be taken as an absolute rule, it is a common pattern. Thus, the bands, which we would expect to be strong, are due to the more symmetric vibrations.

Table 1.1. Single vibration and group frequencies and possible intensities of peaks commonly identified in Raman scattering (3600–2600 cm⁻¹).

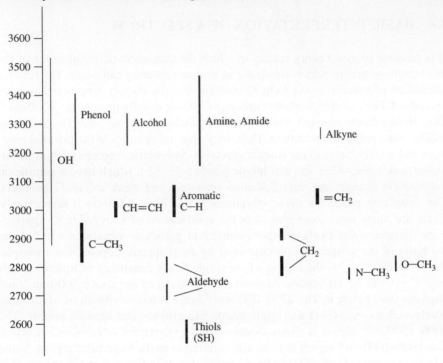

Table 1.2. Single vibration and group frequencies and possible intensities of peaks commonly identified in Raman scattering (2600 to 1700 cm⁻¹).

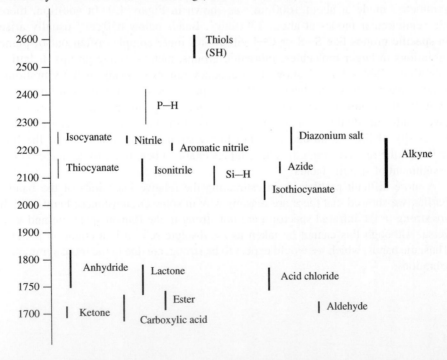

Table 1.3. Single vibration and group frequencies and possible intensities of peaks commonly identified in Raman scattering (1700 to 1200 cm⁻¹).

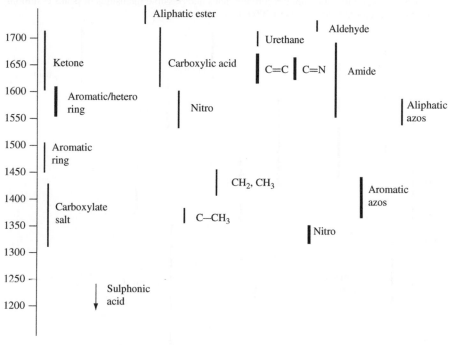

Table 1.4. Single vibration and group frequencies and possible intensities of peaks commonly identified in Raman scattering (1200 to 700 cm⁻¹).

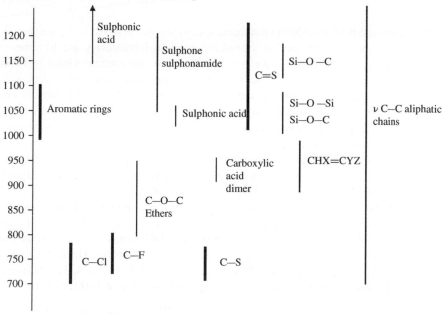

Table 1.5.   Single vibration and group frequencies and possible intensities of peaks commonly identified in Raman scattering (700 to 200 cm^{-1}).

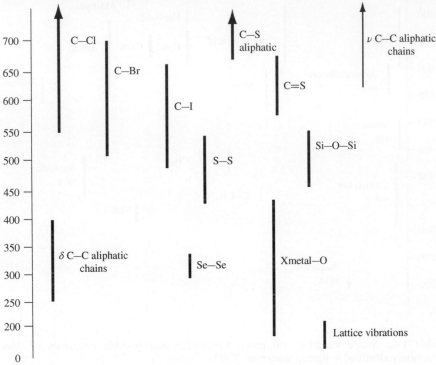

This approach is often used in vibrational spectroscopy. However, to assign specific peaks in the spectrum to specific vibrations, modern laboratories use libraries in which complete spectra are stored electronically. Most spectrometers have software to obtain a computer-generated analysis of the similarities and differences with spectra in these libraries so that specific substances can be identified positively and easily. In other areas, the initial assignment can be confirmed by calculation where the great advantages are a more accurate assessment of the nature of the vibrations and hence of molecular structure.

Some bands in Raman spectra are environmentally sensitive bands, e.g. OH, NH, and are broad and weak. By contrast some backbone structural bands in organic molecules tend to be strong and sharp. The extent of this difference can be illustrated from the fact that water can be used as a solvent to obtain the Raman spectra of organic molecules. This indicates the strength of aromatic bands in the organic molecule and the weakness of hydrogen-bonded OH bands in water. It is this greater selectivity which leads to the simplicity of the spectrum. Thus, the Raman spectra of quite large molecules show clear bands. In Figure 1.4, for example, the IR spectrum is complex and has a strong band just above 1600 cm^{-1} from the carbonyl group due to the C=O vibration. The strong bands in the Raman spectrum are largely due to the aromatic group. The band at $2900\,\mathrm{cm}^{-1}$ due to the CH_2 group is hidden under the strong OH bands in the IR spectrum but can be clearly seen in the Raman spectrum.

The above information makes it possible to start to assign and interpret Raman spectra. The phrase 'interpretation of Raman spectra' is used in many different ways. The spectrum of a molecule can be the subject of a full mathematical interpretation in which every band is carefully assigned or of a cursory look to recognise a pattern of frequencies and intensities and produce the interpretation 'Yes that is toluene'. The libraries in a modern instrument may make identification of a molecule from the spectrum simple in some circumstances but this is only as good as the information in the library and the quality of the sample. Most spectroscopists will still carry out an interpretation to increase the certainty of the final result. The way to do this is described in detail in the next chapter but the most important point is to make sure that as much as possible is known about the sample. The type of compound likely to be present is often known and if possible it is always good to run an infrared spectrum for comparison.

Figure 1.10 is an example where information from this chapter was put to good use in pattern matching. It shows the spectrum of aspirin taken from a tablet with no preparation, from a similar tablet in a plastic bag and a sample of aspirin synthesised by an undergraduate and retrieved before the sample was destroyed. One glance shows that all three are basically the same compound. Where the likely compounds are known, Raman spectroscopy can often give a very specific footprint of a compound and this can be made more certain by accurately comparing the frequencies and intensities with a known pure sample. Actually the first student sample analysed was very different from that in Figure 1.10. The samples were waiting to be safely destroyed and the first sample supplied was indomethacin synthesised as part of the same experiment. This was identified by us immediately the spectrum from the powder which was run without any sample preparation, was available. Bands at about 1600, 1300 and $1000\,\mathrm{cm}^{-1}$ were clearly from the phenyl ring. Had the spectrum been extended further, bands below and above $3000\,\mathrm{cm}^{-1}$ would have been recorded from the aliphatic CH_3 group and the phenyl ring, respectively. Closer examination shows that there is a weak band, arrowed in the commercial sample, which was not present in the student sample. This is present at about the frequency expected for carbonate. It almost certainly comes from the filler in the tablet. Since Raman intensities vary considerably, it cannot be assumed that this is a minor component. Now look again at the student sample between 1600 and $1500\,\mathrm{cm}^{-1}$. There are weak bands here not present in the commercial sample indicating some impurities. With the same argument as before, the fact they are weak does not guarantee that there is not a significant amount of material present.

Raman spectroscopists have to make a number of choices in deciding how to examine a sample and the type of answer required may ultimately determine these choices. If a reference spectrum is not available then band-by-band interpretation is required.

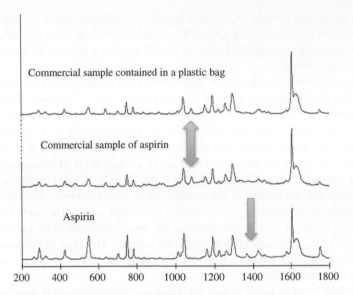

Figure 1.10. Comparison of aspirin spectra with impurity bands arrowed.

Interpretation can be much improved with knowledge of the structure to narrow down the possible bands present and experience can also provide additional assurance as in the case of the NH stretch discussed below. The spectrum of paracetamol shown in figure 1.11 was subjected to structural interpretation rather than pattern matching. Beginning at the high wavenumber end, the band in the 3500–3200 cm^{-1} region could be OH or NH but NH is often towards the higher energy end of the range and from the shape it is most likely to be NH. Bands just above 3000 cm^{-1} are due to H attached to a double bond, whilst those below are due to aliphatic CH. The 1640 and 1540 cm^{-1} bands are due to the carbonyl stretch associated with an amide group (amide I band) and the amide II band, respectively, whilst the 1600 cm^{-1} band is typical of aromatic rings. Note the relative strengths of aromatic band to the aliphatic and amide I band. Bands below 1500 cm^{-1} are due to the backbone of the molecule and the environment to which it is in. These are the fingerprint bands discussed earlier. They are a large number of them and they are dependent on the total physical and chemical structure of the molecule and the environment in which the molecule is in when the spectrum is measured. The pattern can be used to help identify the molecule.

The simplicity and flexibility of Raman scattering have considerable advantages but if care is not taken in making the correct choices, poor or spurious results can be obtained. Chapter 2 describes the choices and provides the background information to enable the recording and interpretation of Raman scattering in a reliable and secure manner.

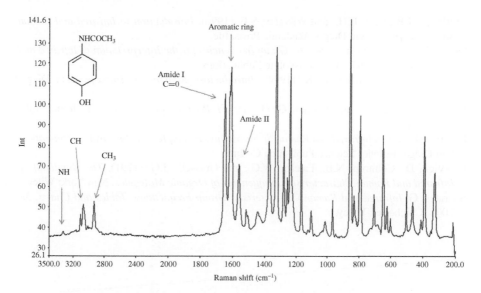

Figure 1.11. Paracetamol spectrum with significant chemical groups.

1.7 SUMMARY

In this chapter, we have attempted to introduce the reader to the basic principles of Raman spectroscopy, without going in to the theory and details of practice too deeply. Chapter 2 outlines the practical choices to be made in carrying out the Raman experiment in full. Later chapters give the theoretical background required for full analysis of spectra, a guide to ways in which Raman spectroscopy has been successfully employed and lead to the more sophisticated but less common techniques available to the Raman spectroscopist.

REFERENCES

1. Smekal, A. (1923). *Naturwissenschaften* **43**: 873.
2. Raman, C.V. and Krishnan, K.S. (1928). *Nature* **121**: 501.
3. McGrane, D., Moore, S., David, S. et al. (2014). *Appl. Spectrosc.* **68** (11): 1279–1288.
4. Fadini, A. and Schnepel, F.-M. (1989). *Vibrational Spectroscopy: Methods and Applications*. Chichester: Ellis Horwood Ltd.

FURTHER READING

Listed here are a number of publications which the authors have found useful for reference, for theoretical aspects of the spectroscopy and for aids in interpretation.

Adams, D.M. (1967). *Metal – Ligands and Related Vibrations*. London: Edward Arnold Ltd.

Colthrup, N.B., Daly, L.H., and Wiberley, S.E. (1990). *Introduction to Infrared and Raman Spectroscopy*, 3e. San Diego: Academic Press, Inc.

Degen, I.A. *Tables of Characteristic Group frequencies for the Interpretation of Infrared and Raman Spectra*, 1997. Harrow: Acolyte Publications.

Ferraro, J.R. and Nakamoto, K. (1994). *Introductory Raman Spectroscopy*. San Diego: Academic Press.

Hendra, P., Jones, C., and Warnes, G. (1991). *FT Raman Spectroscopy*. Chichester: Ellis Horwood Ltd.

Larkin, P.J. (2011). *Infrared and Raman Spectroscopy: Principles and Spectral Interpretation*. San Diego: Elsevier Science Publishing Co Inc.

Lin-Vien, D., Colthrup, N.B., Fateley, W.G., and Grasselli, J.G. (1991). *The Handbook of Infrared and Raman Characteristic Frequencies of Organic Molecules*. New York: Wiley.

Socrates, G. *Infrared and Raman Characteristic Group Frequencies, Tables and Charts*, 3e, 2001. Chichester: Wiley.

Chapter 2

The Raman Experiment – Raman Instrumentation, Sample Presentation, Data Handling and Practical Aspects of Interpretation

2.1 INTRODUCTION

The Raman spectroscopist has to make a number of choices in deciding how to examine a sample and the choices made are ultimately determined by the availability of equipment and by the type of answer required. Should the excitation source be in the UV, visible or near-infrared (NIR) frequency region? Should the detection system consist of a dispersive monochromator with a charge coupled device (CCD) detector or a system using a Fourier transform (FT) and an interferometer with an indium gallium arsenide (InGaAs) detector? Are suitable accessories available to allow the sample to be studied efficiently? How should the sample be presented to the instrument and how can photodegradation and fluorescence be avoided? How do these choices affect the answer? How can the data be interpreted most effectively? This chapter describes the common types of spectrometers which are used, the accessories available for these instruments, the way in which the sample is presented to the instrument and the way to use data manipulation effectively. Finally, there is a guide to how best approach the interpretation of the spectra. The intention here is to give guidance in the thought processes required to answer the above questions and assess more specialist articles effectively.

Modern Raman Spectroscopy: A Practical Approach, Second Edition. Ewen Smith and Geoffrey Dent.
© 2019 John Wiley & Sons Ltd. Published 2019 by John Wiley & Sons Ltd.

2.2   CHOICE OF INSTRUMENT

In Chapter 3, when the theory is developed, it will be shown that the intensity of the scattering is related to the power of the laser used to excite the scattering, the square of the polarizability of the molecule analysed and the fourth power of the frequency chosen for the exciting laser. Thus, there is one molecular property, the polarizability, from which the molecular information will be derived and there are two instrumentation parameters, which can be chosen by the spectroscopist. This choice is not straightforward. For example, since the scattering depends on the fourth power of the frequency, the obvious way of improving Raman sensitivity is to use the highest frequency possible, which would mean working in the UV region. UV excitation also has the advantage that there is little or no fluorescence in the energy range covered by Raman scattering. However, this is not the most common choice. Many compounds absorb UV radiation and the high energy of the photons in this region means that there is a high risk of sample degradation through photodecomposition and burning. It also means that the spectra may differ from normal Raman spectra due to resonance with any electronic transition, which may cause absorption. This changes the relative intensities of the bands (see Chapter 4 for an explanation of resonance). Additionally, the lasers used can be quite expensive, there is a problem with safety since the beam is invisible and the quality of the optics required in the UV is higher than for visible and infrared radiation. However, the rapid development of optical devices including laser diodes, which work in the blue or UV, the unique information that can be obtained in this region and the improved sampling methods now available mean that UV Raman scattering is becoming more widely used. An example of the use of UV Raman spectroscopy is given in Chapter 7.

Currently, laboratories choose either from a range of dispersive spectrometers with differing laser line sources or an interferometer with FT software. Both types of instrument have their advantages and disadvantages. Historically, FT systems were preferred for use with infrared sources because the detectors were better than for dispersive systems but now good detectors are available for dispersive systems. The choice for a particular laboratory very largely depends on the type of analysis to be carried out and the materials which the instrument is expected to examine.

This section outlines the main considerations in choosing a system. They are the choice of laser, the choice of filter or second monochromator, the choice of monochromator or interferometer and the choice of sampling optics. For many users, these choices will have been determined by equipment availability already but this section will make it possible to evaluate the suitability of the equipment available for a specific task. Figure 2.1 shows a schematic of the main parts of a dispersive Raman spectrometer, each part of which plays a key part in the performance of the instrument and each of the areas numbered in red will be considered in this section.

The green arrows indicate the direction of the laser beam and the scattered light up to the point where the Raman scattered light is separated from the stronger Rayleigh scattered light. The incident laser beam is directed onto an interference filter at an angle

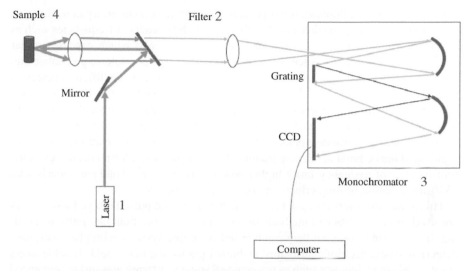

Figure 2.1. A simple diagram outlining the basic components of a dispersive Raman spectrometer. The numbers in red correspond to the four areas described in the text.

that reflects it off the surface and through focusing optics onto the sample. In many instruments, the beam diameter will have been expanded to fill the optic elements and to prevent damage to the filter. Above the sample, the scattered radiation is collected from as wide an angle as possible and sent back to the filter. The scattered light hits the filter at a different angle to the incident laser beam and passes through it. The filter is constructed to remove radiation at the frequency of the laser radiation allowing only frequency-shifted light to pass through. Note that the colour changes shown are greatly exaggerated for clarity. Raman-shifted light would only appear as a shade change. In addition, the optics shown are crude. The radiation is then focused into a monochromator where a grating splits it up into different frequencies, which are detected on a CCD device and sent to the computer to be processed.

The choice of laser (area 1 in Figure 2.1) very much depends on use. The use of UV sources is less common for the reasons already outlined but cheaper, reliable UV sources are now available and UV scattering can give unique information, so for particular uses more of these systems will appear (see Chapter 7). For the more commonly chosen infrared and visible regions the basic criteria are that the power and frequency are stable, the lifetime is long and the frequency band width is narrow. It used to be that visible lasers such as argon or krypton gas lasers were preferred but cheaper, stable diode lasers and solid-state lasers made with materials such as Neodymium-doped yttrium aluminium garnet (NeYAG) used either as is or frequency doubled are now more common.

The main problem with visible lasers is that many compounds fluoresce in the visible region. Since Raman scattering is relatively weak compared to fluorescence in many cases where fluorescence is present, the analyte or an impurity may fluoresce sufficiently to swamp the detector. For this reason, NIR lasers at frequencies such as 795 or 785 nm where fewer compounds will fluoresce and which can use similar detectors are a common choice. As discussed below, it is possible to go further into the infrared and so further reduce fluorescence interference using different detection systems discussed below. NeYAG lasers give a very stable beam at 1064 nm, which is widely used and a good choice if a wide range of samples or difficult samples are to be analysed. 1280 nm is now becoming more common and even 1550 nm sources are used. One big advantage is that for some samples there is little adsorption in this region and hence good depth penetration. However, since scattering decreases as the fourth power of frequency, much higher powers have to be used and particularly with 1550 nm radiation heating effects can be a serious problem.

This leaves two more types of lasers, tunable lasers and pulsed lasers. Pulsed lasers are used in more advanced methods often to involve more than one pulse in exciting the scattering event but they can be used in simple systems with phase-sensitive detection to help discriminate against ambient light for use in the field. Tunable lasers are valuable in techniques such as resonance Raman scattering and surface enhanced Raman scattering where scattering efficiency does not follow the fourth power rule but is dependent on the electronic spectrum of the analyte. Examples of the use of these are given in Chapters 4–7.

In Figure 2.1, one filter (area 2) is shown as removing all non-frequency shifted radiation. Of course, this is not 100% efficient and to increase efficiency, many commercial instruments use a set of filters at this point. There are two basic types, notch filters which remove radiation at the frequency of the laser and in a region on either side of it (commonly about 200 cm^{-1} on both sides but it can be less – Figure 2.2) and edge filters which remove all light above a certain frequency so that only Stokes scattering and any other lower energy light due to fluorescence, etc. is transmitted. Clearly this prevents anti-Stokes scattering being recorded but edge filters have the advantage of being cheaper with a long life. As Figure 2.2 shows, it would be possible to measure a Raman spectrum without a filter in place but in these circumstances a large amount of non-frequency shifted light is present in the monochromator and some is picked up by the detector reducing its efficiency, or completely swamping it or even damaging it.

The disadvantage of these filters is that they restrict the user to one excitation wavelength and if the sample absorbs or fluoresces at that wavelength, the results can be poor. For this reason, some instruments have arrangements that allow filter changes so that, for example, 532, 633 and 785 nm excitation could be used in turn to cover enough range so that scattering from most samples can be obtained and more expensive tunable filters are available. However, filters are not the only way of discriminating Raman scattering from Rayleigh scattering. Most of the older spectrometers used two or even three monochromators linked together and this approach is used in some modern systems to give flexibility. Essentially the main function of the first monochromator is to remove much of the non-frequency shifted light. In some studies

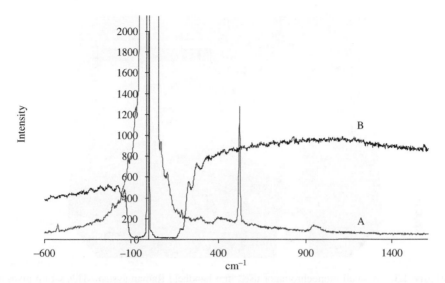

Figure 2.2. Raman spectra taken across the exciting line. In A the intensity of the Raman band at 520 cm⁻¹ is much weaker than that from the non-frequency shifted light. The second spectrum B is from a poor Raman scatterer with a notch filter in place. No Raman bands are observed but there is some nonspecific radiation, possibly weak fluorescence, which causes a signal. Close to the exciting line this radiation is removed by the notch filter with some laser breakthrough at the laser energy. The features on the edge of the region covered by the filter are artefacts caused by the filter and not Raman peaks.

where the effect of excitation frequency on wavelength is of importance, a tunable laser, with either a tunable filter and single monochromator or a double monochromator may be best since the frequency to be rejected will vary with excitation frequency. However, compared to a double monochromator system, notch and edge filters make for smaller and simpler efficient instruments and so are widely used.

Most instruments have other filters such as neutral density filters near the laser to help control laser power and a filter to remove any radiation not at the laser frequency such as laser sidebands. Other elements such as those required for polarization measurements are discussed later.

Figure 2.1 shows a schematic of a Czerny Turner monochromator (area 3) used to separate the radiation into different frequencies and a CCD device to detect each frequency. The grating in the monochromator is used to disperse the radiation which is then focused to form a sharp image on the CCD, which spans the full width of the CCD. A CCD detector is a sectored piece of silicon in which each sector is separately addressed to the computer. In this way, it is possible to discriminate each frequency of the scattered light and therefore construct a spectrum of the type shown in Chapter 1. In the past, the monochromators used to remove the fluorescence could be quite large and up to a meter in length to give the resolution required to accurately define the sharp Raman bands. However, the use of optics

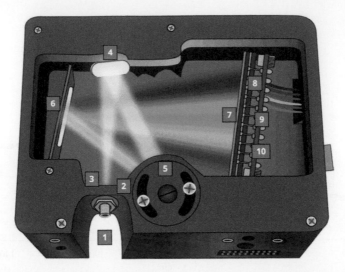

Figure 2.3. A small monochromator used in a handheld Raman system. This set up gives a longer distance between the grating and the CCD compared to the arrangement in Figure 2.1 increasing the dispersion possible in a small unit. 1–3 entry, 4 mirror, 5 grating mount, 6 focusing mirror and 7–10 detector. Source: Image reproduced by courtesy of Ocean Optics.

which sample small volumes and give a small image, diode lasers, filters and small detectors mean that a Raman spectrometer can be quite compact. To aid this, more space-effective arrangements in the monochromator than that shown in Figure 2.1 are now widely used, an example of which is shown in Figure 2.3. The device shown is small because as well as the monochromator design, the image at the entrance is small allowing a small grating and detector to be used. However, there is a limit to the resolution which can be obtained because of the short path length between the grating and the detector.

The grating which is chosen to disperse the radiation will depend on the blaze wavelength, i.e. the wavelength at which scattering is most efficient, and on the number of lines per centimetre. The more lines the greater the dispersion. The detector is usually a CCD or CMOS chip. This is a sectored piece of silicon and the radiation from the grating requires to be accurately focused onto the surface, otherwise sensitivity is compromised. The sharp nature of Raman bands means that it is preferable to use a grating with many lines to obtain a wider dispersion and hence the higher resolution required to define them. However, since only the radiation falling on the CCD is detected, the choice of grating and CCD has to be matched. A broader range of frequencies can be detected by the CCD if a grating with a lower dispersion is used but the resolution of the bands may be compromised. To overcome this some instruments use a high-resolution grating and a narrow spectral range and move the grating to detect different ranges, which can be combined by software to get the full spectrum.

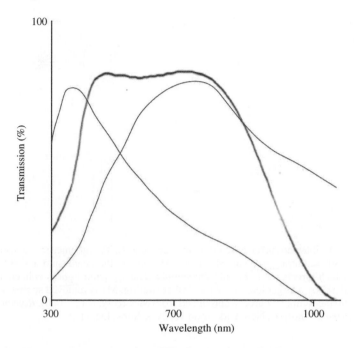

Figure 2.4. Wavelength response of a CCD (blue) suitable for Raman scattering and two gratings (red) blazed at different wavelengths.

Figure 2.4 shows the wavelength response of a grating and CCD suitable for a system with visible excitation. It is obvious that the efficiency of the system will vary with wavelength. The instrument will be calibrated to compensate for this, so the user does not usually need to worry about it. However, if the ends of the ranges are used as is often the case when using a CCD detector with NIR radiation, the high-frequency bands in samples either with high backgrounds or which give weak scattering can appear as having lower intensities than should be the case.

Gratings and filters can be changed to move further into the infrared but CCD detectors are limited in this respect so that the longest wavelength excitation commonly used is 830 nm. To reduce fluorescence to an absolute minimum, excitation wavelengths such as 1064 and 1280 nm or longer are desirable. However, because of the fourth power law, scattering efficiency gets progressively lower the further into the infrared the excitation frequency. This leads to higher powers being required, which can cause sample heating so the decision on excitation frequency depends on the application. Dispersive instruments with excitation in the range from 830 to 1550 nm use InGaAs detectors in an analogous way to CCD detectors and this is effective. However, there is an alternative. With a technology similar to that used in infrared spectrometers, effective systems can be built using

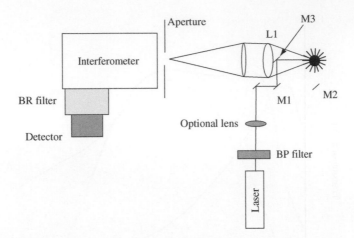

Figure 2.5. An interferometer-based Raman spectrometer. Two geometries are usually used. As shown, 180° scattering is obtained. Mirror M1 directs the beam onto a small mirror or prism M3, which directs the light onto the sample. The scattered light is collected from the region round M3 by a larger lens. To obtain 90° scattering, M1 is removed so that M2 reflects the beam onto the sample. Source: Reproduced from McCreery, R.L. (2000). *Raman Spectroscopy for Chemical Analysis*. New York: John Wiley & Sons, Inc. [1].

an interferometer instead of a monochromator, an InGaAs detector and usually a 1064 nm laser. In these systems, the monochromator is replaced with an interferometer (Figure 2.5). In an FT system, all scattered radiation without frequency separation is divided by a beam splitter into two beams, which are reflected back, one from a fixed and one from a moving mirror, and the two recombined. A detector measures total intensity as the path length of the beam reflected from the moving mirror varies giving an interference pattern. The total intensity and the path difference are then transformed using a Fourier transform (FT) to give a standard spectrum showing the intensities and frequencies of specific peaks. Initially these systems working further into the infrared offered more effective discrimination against fluorescence than a dispersive system with 785 nm excitation and CCD detector. However, now that dispersive systems are available with good IR detectors, the choice is more difficult.

There are two basic geometries used in collecting Raman scattering, 180° scattering often called backscattering and 90° scattering. Figure 2.1 shows a 180° arrangement. The potential for either arrangement is shown in Figure 2.5 and there is no notch filter. Both are effective but most modern systems use 180° scattering.

With the increased flexibility of modern optics there are many more ways the scattered radiation can be collected and many instrument geometries. For example, radiation can be collected using a mirror system such as a Cassegranian system, or a silvered sphere to collect from a larger angle and 'grazing incidence' in which the laser beam is directed along the surface can be used for thin samples. Instruments include handheld systems powered by two A3 batteries, high-powered microscopes and fibre-optic probe systems.

2.3   TRANSMISSION RAMAN SCATTERING AND SPATIALLY OFFSET RAMAN SCATTERING

Careful analysis of the events that occur when Raman scattering arises from a powder, a turbid solution or a particle suspension has led to the development and increased application of transmission Raman scattering (TRS) and spatially offset Raman scattering (SORS). It is simple to take the spectrum of colourless pharmaceuticals such as aspirin and indomethacin. Usually this would be done using 180° backscattering and focusing the beam just into the top of a tablet. However, what if the tablet is not completely homogeneous? This would not be a true representative result of the tablet composition. Up to a certain thickness it is possible to pass the exciting radiation through the sample and collect on the other side (TRS). As the photons penetrate the sample, they will be scattered many times, following a random path through the material. This may take them outside the collection volume for the backscattered radiation but they are still there. At each scattering event, there is a chance that the photon is annihilated and a new Raman-scattered photon created. Initially the direction of the incident photons is straight through the sample so that for a number of events, a significant fraction penetrates downwards with the percentage decreasing with distance. These photons can be collected at the far side of the sample. In a 4 mm thick tablet there are usually sufficient photons to obtain an effective Raman spectrum from the bulk of the sample. This is TRS and it is used in pharmaceutical analysis.

Raman-scattered photons decay at a different rate to elastically scattered photons (called Tyndall scattering in particles instead of Rayleigh scattering). Some photons which backscatter will emerge eventually after multiple scattering events from deep in the sample a significant distance away from the irradiation point. The ratio of Tyndall to Raman-scattered photons favours Raman scattering. By collecting the light away from the point where the exciting beam penetrates the surface, a significant amount of the strong backscattering from Tyndall scattering, reflection from the surface and fluorescence is not collected making detection of the Raman scattering more effective (Figure 2.6). This is SORS. There are a number of possible arrangements for collection of SORS but one obvious one is to collect the scattering by using a ring of fibres round a centre fibre down which the exciting radiation passes as described above. There are many possible uses for SORS. One high-profile use is the detection of illicit materials in opaque objects such as a tube of toothpaste. This has led to the development of SORS for airport security for medical essentials like baby milk.

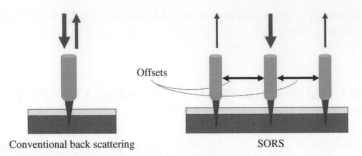

Conventional back scattering   SORS

Figure 2.6. Schematic diagram of spatially offset Raman spectroscopy (SORS) showing a central optic fibre used to deliver the exciting radiation and two of collection ring of fibres used to collect the scattered photons. The offset can be varied to maximize collection of Raman scattering and minimize collection of non-frequency shifted radiation. The conventional scattered radiation is shown as more intense by the thickness of the arrow because it is contains more non-frequency shifted light.

2.4 RAMAN SAMPLE PREPARATION AND HANDLING

Given the many ways in which a sample may be presented to a spectrometer some thought needs to be given as to how that might affect the resulting spectrum. In bench top and handheld spectrometers, Raman spectroscopy is well known for the minimum sample handling and preparation that is required. Hendra's rubber duck [2] is a typical example. A small, children's duck thought to be made of rubber was placed directly in the spectrometer beam. Almost immediately a Raman spectrum was recorded of polypropylene! Whilst a large range of materials can be examined this way, many samples require some form of preparation and/or mounting in a spectrometer. Typical Raman accessories are powder sample holders, cuvette holders, small liquid sample holders (cf. NMR sample tubes), temperature-controlled blocks and clamps for irregularly shaped objects. There are also specialist cells for rotating solids, vapour cells, reaction cells and variable temperature or pressure cells. For systems using a microscope, a mirror or prism may be provided to allow the beam to be deflected so that larger samples can be measured. If this is not the case, a small adaptor, which enables collection from a larger volume can be screwed into the microscope instead of the objective lens. This has a mirror which turns the beam 90° to the microscope direction. A simple holder for the sample such as a 1 cm cuvette or more complex devices such as a spinning sample holder can be mounted on the edge of the microscope stage.

In this section the advantages, disadvantages and precautions required for several ways of handling and mounting samples are described. A review article by Bowie et al. [3] highlighted some of the effects on FT Raman spectra, which can originate from the sample. This section gives some examples of how to overcome the more common effects seen with a range of instruments.

Many organic and inorganic materials are suitable for Raman spectroscopic analysis. These can be solids including polymers, liquids or gases. The majority of bulk industrial laboratory samples are powders or liquids, which can be examined directly by Raman spectroscopy at room temperature. Sample presentation is rarely an issue for bulk samples and many materials can be mounted directly in the beam as neat powders, polymer films, etc. For example, the authors have examined many liquids and powders presented in glass containers from capillary tubes, through vials, to 500 ml brown bottles. Samples have also been examined in polymer containers.

2.4.1 Sample Mounting – Optical Considerations

Sample preparation and mounting can be relatively simple and flexible. However, if reproducible spectra or quantitation is required, then the beam shape and sample presentation needs to be considered. The beam is usually focused into the sample to create a volume where there is a high-power density so that scattering is more efficient and most collected scattering will be from that volume, the sampling volume. The effect of the material on the beam can be quite dramatic and photon migration on depth penetration with turbid or opaque samples has already been discussed. However, in a clear solution the sampling volume will be approximately as shown in Figure 2.7. It can be calculated by choosing a threshold power density below which appreciable Raman scattering is not expected. The depth of this volume is the distance between the converging beam above the sample with that power density and the diverging beam below the sample with the same power density. The diameter is the diameter of the beam at those points. Clearly, there is no sharp cutoff point and some scattering may be expected from above, below and outside this volume and in addition the power density increases as the beam narrows towards the midpoint.

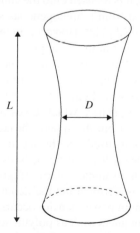

Figure 2.7.   Sampling volume for a simple case. One solution gave $- D = 4\lambda f/\pi d$; $L = 16\lambda f^2/\pi d^2$; D, diameter of cylinder; L, length of cylinder; λ, laser wavelength; d, diameter of unfocused laser beam; f, focal length of focusing lens.

Usually only a rough guide is required so the volume is usually approximated to a cylinder. There are other problems. The beam usually passes through air into the liquid causing refraction and a change in shape of the sample volume.

In coloured solutions and solids, absorption of the excitation radiation causes both attenuation of the beam through the sample volume and local heating due to absorption, which in solution, can cause local changes in the dielectric constant and hence lensing of the beam. One particular problem is that although the exciting beam may penetrate to some depth, the weaker scattered radiation will be absorbed and so that only scattering from near the surface may be detected. This is called self-absorption. As already stated in turbid solutions appreciable photon migration will occur. These matters will be dealt with later in the book for samples where they are important but for many solutions the approximation shown in Figure 2.7 is useful.

This volume can be used to estimate approximately the number of molecules being interrogated in the system at any one time. In the case of a gas, the number of molecules will be quite low. As a result, for gas-phase measurements, instead of focusing the beam tightly, very long cells are often used. Sensitivity can be increased by using concave mirrors to reflect the exciting and scattered beams back through the sample and in some gas cells, a multiple pass system reflects the beam many times to increase sensitivity. Similar arrangements can be made for liquids but these are usually less necessary due to the higher concentration of molecules in the same volume. To increase sensitivity some gas cells have been specifically designed to use light pipes or high pressure.

For homogeneous solids and liquids, collection should be from as large a cone of scattering as possible. For thin samples sensitivity can be improved by using a reflective surface behind the sample tube/vessel. For an FT spectrometer, radiation is passed through the Jacquinot stop, which can have a diameter of a few millimetres, and for a visible spectrometer, it is focused onto the entry slits of the monochromator. To obtain the maximum signal from a homogeneous solid, the surface of the solid should be at, or close to, the focal point. However, in many cases, acceptable spectra can be obtained from samples not at the focal point. It has been shown that relative band strengths can vary depending on the distance the sample is mounted from this point. Whilst this is not always significant in identification, quantification could be seriously affected [4]. For liquids and gases, the system can be improved further by creating a completely reflecting sphere which gives multiple reflections inside the surface and allows only egress from the sphere through a cone, which is collected directly by a lens or is directly focused into the spectrometer.

For qualitative studies, since a focused beam is usually used, in clear solutions most scattering occurs from the sample volume (Figure 2.7). The exact shape and position of the beam is often not critical but the depth into the sample at which the volume containing the maximum power density occurs should be chosen to discriminate against signals from containers such as polythene bottles. Some of these materials give very strong scattering so that even the weaker excitation outside the defined sample volume can create sufficient scattering for signals to be detected, so it is good practice to take the spectrum of the container for reference. For quantitative studies in clear solutions the sample should be set up to collect scattering from as much of the sample volume as is practical.

For colourless powders or coloured powders at excitation wavelengths where absorption and emission are sufficiently low, qualitative spectra are again easy to obtain. However, the combination of photon migration and refraction, lensing and heating, and absorption of the scattered light make quantitation difficult. It is not impossible particularly with very strong scatterers since quite dilute samples can be used. Even in qualitative studies, the collection solid angle has to be considered. On some occasions such as when crystalline samples are examined, the angle of the sample to the scattered beam, i.e. 90° or 180°, will lead to orientation effects. Rotating the sample in the beam can average out these effects. Particle size effects have also been reported.

A common problem with samples for Raman spectroscopy arises from fluorescence interference. Most modern instruments have very good background subtraction routines and since fluorescence bands are normally much broader than Raman bands, fluorescent backgrounds can often be subtracted out. Where multiple excitation wavelengths are available, one can be chosen so that it lies either above or below the wavelength of the fluorescence band, thus separating the Raman scattering from the fluorescence. In some cases, where the fluorescence arises from an impurity it can be burnt out by leaving the sample in the beam for a few minutes or overnight. This works because there is specific absorption of the radiation into the fluorophore so that it is preferentially degraded. However, particularly if the sample itself is coloured, absorption can cause degradation. Some modern instruments have been built to remove fluorescence optically. They take two spectra using closely spaced excitation wavelengths. The small shift in peak positions separates the sharp Raman bands but the broad fluorescence is little affected. Subtraction of the two leads to clear discrimination of the Raman spectra. There are other ways of overcoming fluorescence such as adding a quencher or adsorbing the sample as a thin film on a metal surface as is done in surface enhanced Raman scattering (Chapter 5).

The intrinsic intensity of Raman scattering per molecule (Raman cross-section) varies widely between materials and consequently, the differing scattering intensities of the analyte and the surrounding matrix need to be considered, as does the possibility of contamination. If the container or an impurity has a high cross-section (or is resonant, Chapter 4) then scattering from that source can dominate or be an appreciable factor in the spectrum and can lead to false assignments. There are a number of examples in the literature where this simple precaution has been ignored and important conclusions drawn from data, which subsequently has been shown to have arisen from a contaminant. An empty polythene bottle placed in the beam will show bands due to polythene. Fill the bottle with sulphur and only the sulphur bands will be observed as the polythene is a much weaker Raman scatterer. Water is a strong absorber of infrared radiation, as is glass, but both are weak scatterers in Raman spectroscopy, which makes the technique particularly suitable for samples in aqueous solutions and/or in glass containers. However, glass and water do have their own spectra and need to be considered with weak solutions containing weak scatterers.

Small samples may have to be examined with a microscope or microprobe but this means that the beam diameter reduces very significantly and is often much smaller than the total size of the sample. The focal point will then determine which part of the sample is being analysed. This means that particularly if a larger sample is used it is important to check the homogeneity of the sample by taking a number of measurements across it. The relative refractive indices of the sample and matrix may also have an effect. This is important when attempting confocal Raman microscopy and will be dealt with in Section 2.7.

2.4.2 Raman Sample Handling

As already stated, powders and liquids can often be examined by placing the container in which they are supplied directly in the beam. The only constraints are that the outside of the container be clean, free from fingerprints, which cause fluorescence, and the labels do not obscure the sample although this restriction can be less important if SORS is used. It is important that the laser is focused into the sample and away from the walls of the cuvette or the sides and foot of the microtiter plate. When this is done, the significantly higher power density at the sample mitigates against interference from the cuvettes or microtiter well plates. However, if the beam is focused onto these materials, excellent spectra can often be obtained from the polymer. This is particularly important for plastic well plates used to run large numbers of samples since the volume of the wells can be quite small so that the distance between the sample volume and the wall may not be sufficient to prevent some bands from the polymer appearing in the spectrum.

Neat powders with weak signals can be mounted in loosely filled containers or in a compacted solids holder. The authors used the latter technique successfully with a crystalline, low-density fungicide which was moved away from the bottle wall by the laser beam power, but gave a strong spectrum 'fixed' in the holder. However, with samples that are crystalline, orientation effects can change the spectra as can the particle size of powders. Using inorganic material it has been shown that the Raman intensity increases as the particle size decreases [5–7]. The theoretical dependence has been described by Schrader and Bergmann [8]. Experimental work has shown that a general fit can be obtained. However, if a sample is dispersed in a matrix, e.g. filler in a polymer or paint resin, droplets in an emulsion, then a sudden and rapid reduction Raman signal can occur at particular sizes below the wavelength of illumination. An example of this is titanium dioxide which gives characteristic Raman spectra in the bulk solid state but gives weak, or no spectra, when dispersed as a filler in polythene.

Photodegradation or burning can be minimized by lowering the power and increasing the accumulation time to allow energy to dissipate from the interrogation site. If this is not enough, either changing the excitation frequency to one which absorbs less or moving the beam or sample during acquisition so that for any one spot on the sample excited states decay and heat is dissipated before the spot is excited again. It would seem obvious that, if available, moving to an infrared frequency where there is less absorption would be effective and in many cases it is the best solution. However, the fourth power law makes scattering less efficient leading to the use of higher excitation powers and absorption though infrared overtones can occur so that heating can become a problem, so care is required particularly with 1550 nm excitation. If an excitation wavelength shift is not possible or desirable, moving the beam provides a good solution and many spectrometers provide a rastering system, mainly for mapping, in which the beam is moved across or round a sample. Homogeneity of the sample needs to be considered. This method will give an average of the spectrum over the whole area scanned. This may or may not be advantageous depending on the information required. If it is not, mapping or imaging is most likely to be appropriate as described later. Alternately the sample can be moved. For solutions, either a flow through cell or a spinning cell can be used. Powders can be packed into a channel on a disc or compacted into a disc itself. The laser is focused onto the sample away from the centre of the disk so that as the disc spins scattering is obtained from different parts of the sample allowing time for excited states to decay and heat to disperse before the same part of the sample is interrogated again. The speed of rotation has to be kept <50 Hz in FT Raman spectrometers or beats may be seen across the spectrum. An alternative way to reduce the burning effects in solids is to disperse the sample in other media without a Raman spectrum such as KBr or KCl. Strong spectra have been recorded at high laser powers (1400 mw at 1064 nm), without burning, by this method. A study of the various diluents concluded that KCl was often the best diluent [9]. The process of forming the disks requires pressure and can cause changes in the sample; consequently, this method should be avoided if physical forms such as polymorphism are to be studied. Accessories are available which will turn samples at a speed which will prevent beats in the spectrum [10]. The preparation of samples as hydrocarbon oil mulls between salt flats in the same way as for infrared can give good, strong spectra with less risk of burning (Figure 2.8) [12]. This preparation technique also preserves physical form for polymorphism studies.

Although using infrared excitation greatly reduces fluorescence problems, there are some samples which still fluoresce. Copper phthalocyanine (CuPc) is unusual in that not only does the visible Raman spectrum show some fluorescence but the fluorescence is more of a problem with 1064 nm excitation (Figure 2.9). Early work by the authors showed that blue, green, red, yellow, some brown and even some black samples could be examined using 1064 nm excitation, but blues and greens based on CuPc would still fluoresce. On increasing the exciting wavelength to 1339 nm, the fluorescence is much reduced and CuPc bands reappear. It has been suggested [15] that this strange phenomenon is due to the transition metal in the ring being present in the phthalocyanine ring but spectra, recorded with 1064 nm excitation, of various other metal-substituted phthalocyanines, including metal-free phthalocyanine [14] do not show this effect (Figure 2.10).

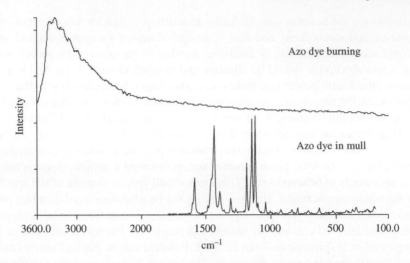

Figure 2.8. Neat sample burning vs. mull spectrum. Source: Reproduced from Chalmers, J.M. and Dent, G. (1997). *Industrial Analysis with Vibrational Spectroscopy*. London: Royal Society of Chemistry [11].

In Figure 2.10, a somewhat better spectrum of CuPc is shown. This was obtained using another method of reducing background by making a disk from a mixture of CuPc in powdered silver (1 : 1000) by compaction using a standard KBr press. The disk was not spun. This preparation disperses the highly adsorbing CuPc as fine particles on a silver surface and enables energy transfer to the silver. It should be noted that repeated use of the press to make these disks can distort the plattens in the press.

Colour is not a good guide as to whether a sample will fluoresce. Some coloured materials adsorb light and lose the energy by non-radiative pathways. Further, clear, water white crystals have been observed to cause fluorescence at all illuminating wavelengths. The reason for this is not known but clear crystals allow significant penetration by the laser so that the sample volume can be quite large, meaning collection from many molecules. Thus, a very low concentration of strongly fluorescent material might be detected.

The occurrence of fluorescence as a problem is much reduced at higher excitation frequencies because fluorescence is emitted from the lowest energy levels and therefore is in a frequency range which is much lower than that, which will be captured to measure Raman scattering. In addition, the fourth power law means that scattering is stronger. However, even at higher excitation frequencies, some solid samples can still exhibit fluorescence, which interferes and often there are more absorption bands increasing problems with burning, self-absorption and photodecomposition. Figure 2.9 shows both Raman scattering and fluorescence in the one spectrum but separated in frequency. If the excitation frequency is moved to a slightly longer wavelength (lower energy), the Raman shift will also move to lower energy but the

(a)

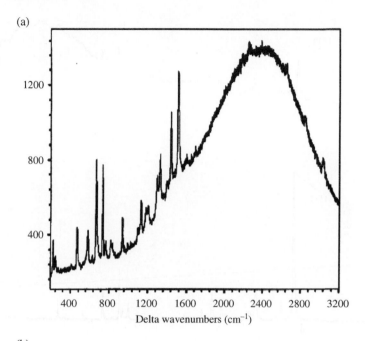

Delta wavenumbers (cm^{-1})

(b)

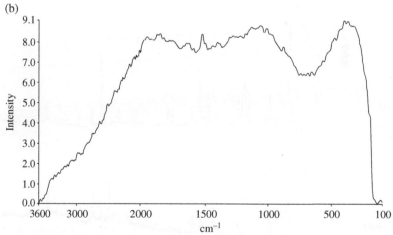

Figure 2.9. Raman spectrum of copper phthalocyanine: (a) With 632 nm excitation. Source: Reproduced from Chalmers, J. and Griffiths, P. (eds) (2001). *Handbook of Vibrational Spectroscopy*, vol. 4, New York: John Wiley & Sons, Inc., pp. 2593–2600 [13]. (b) With 1064 nm excitation. Source: Dent, G. and Farrell, F. (1997). *Spectrochim. Acta* **53A** (1): 21 [14] © 1997 by kind permission of Elsevier Science-NL, Sara Burgerhartstraat 25, 1055 KV Amsterdam, The Netherlands.

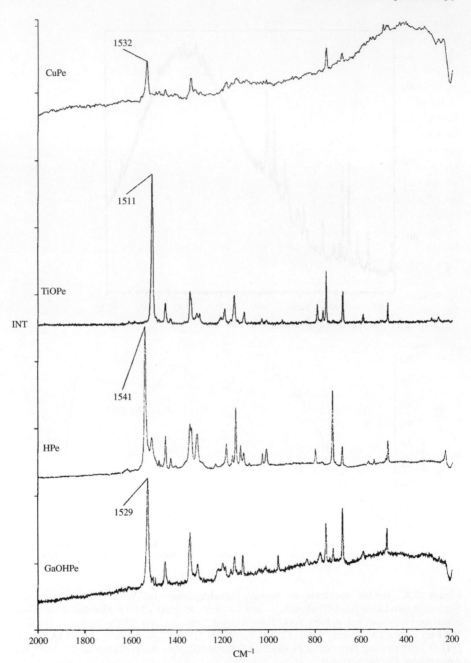

Figure 2.10. NIR FT Raman spectra of phthalocyanines with various metals. Source: Reproduced from NIR FT Raman examination phthalocyanines at 1064 nm, Dent, G. and Farrell, F. (1997). *Spectrochim. Acta* **53A** (1): 21 [14] © 1997 by kind permission of Elsevier Science-NL, Sara Burgerhartstraat 25, 1055 KV Amsterdam, The Netherlands.

fluorescence will not change in energy so the Raman signal will appear on top of the fluorescence and could require significant background correction. Of course, since the fluorescence will then be excited at a different wavelength it could go either up or down in intensity. However, it would obviously be better to choose a much lower or higher excitation frequency.

Liquids are less prone to burning than solids due to the high mobility, which dissipates the heat throughout the solution. If the samples are relatively clear or lightly coloured, spectra can be enhanced by placing the samples in silvered holders, which reflect the signals back through the sample in some cases many times thus increasing the scattering. Only small solid samples or adsorbing liquids can be examined this way because at longer path lengths the scattered radiation will be dissipated by photon migration or be adsorbed by the sample itself (self-absorption). This effect was demonstrated with tetrahydrofuran (THF) where the $917\,cm^{-1}$ band when excited using 1064 nm varied. It has an absolute position of $\sim\!8478\,cm^{-1}$, which is almost at the peak of the NIR absorption band due to the second overtone of the C—H stretch. This results in the strength of the band being attenuated due to self-absorption of the scattered radiation [16, 17].

Polymers of all shapes and sizes can be examined by Raman spectroscopy. Safety spectacles, rolls, thin films, bottles and moulded plattens have been examined (see Section 6.4). Some important classes of polymer are relatively weak scatterers and this can either be regarded as a problem or be used to advantage. For example, as mentioned earlier, sulphur in a polythene bottle gives a very strong Raman spectrum with no evidence of the bottle wall in the spectrum. However, the spectrum of a 2% azo dye in polymer film shows bands due to both dye and polymer (Figure 2.11).

In examining a polymer film one recommendation is to fold the film as many times as possible to create a 'thick' layer. In this case any orientation effects will be lost.

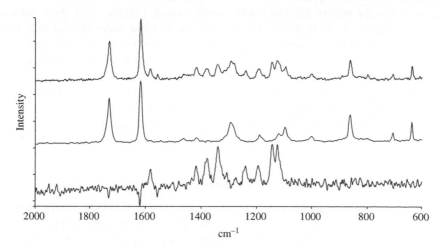

Figure 2.11. Raman spectra of dye in fibre. The top spectrum is from the dyed film, the middle one is from the film and the foot one is the difference.

Sometimes film samples are not big enough to fold. An enhanced, strong spectrum can be recorded by placing a small, single sheet flat across the mirrored back face holder. Spectra of coloured polyethylene terephthalate have been recorded this way with strong enough bands to see both the dye and the film. In Figure 2.11, the spectra of both clear film and dyed film were recorded by this technique. The resulting spectrum from a spectral subtraction clearly shows the bands due to the dye.

2.5   SAMPLE MOUNTING ACCESSORIES

The wide variety of Raman spectrometers now available from handheld systems to microscope systems and relative ease with which Raman scattering can be collected, meaning that there are a very wide range of sampling accessories available. This section deals with sampling when a microscope is not available.

2.5.1   Small Fibres, Films, Liquids and Powders

Many samples that cannot be examined directly can easily be mounted in a simple holder at the optimum point by the use of small-diameter glass tubes. NMR sample tubes are often used for liquids, or loosely packed solids, and are easily held in position. Solids can also be held in the open end of the tube, then mounted so that the beam is focused onto the powder rather than through the glass wall. If the powder in the main part of the tube exhibits thermal degradation, then slowly rotating the tube constantly refreshes the exposed surface. Fibres and thin polymer films can be examined by loosely packing into the tubes or by wrapping round the outside of the tube until a thickness is achieved, which will provide a spectrum with a required *S/N* ratio. Again, if burning occurs, the tube can be slowly rotated. Polymers and fibres can also be examined by wrapping them around a glass microscope slide. Special cells have been designed [18] to examine fibres and fabrics, which involve both sample compression and a backscattering mirror. The cell also contains a windowless aperture. Previously it was stated that powders which are strongly absorbing can be diluted by KCl, KBr, Nujol, etc. These samples can be mounted in glass tubes in the same way as the neat powders, or compressed into disks and mulls as they would be for infrared examination and mounted directly. For microscopes, capillary tubes can be used or microscope slides with powder scattered on them and covered with a cover slip.

2.5.2   Variable Temperature and Pressure Cells

This is an area where a wide range of specifically designed cells have been reported to fulfil the requirement for both dispersive and FT spectrometers (Figure 2.12). Cells have been designed to work across a temperature range of -170 to $950\,°C$ and from high vacuum to $10\,000\,psi$. Controlled temperature cells, which can be set to cycle to carry out specific reactions such as PCR, can be used with little modification,

Figure 2.12. Variable pressure and temperature cells. Source: Images by courtesy of AABSPEC.

for example, by focusing into the top of the wells in a microtiter plate mounted on a stage, which can be shifted manually or automatically with a computer.

Where pressure changes are to be measured, the difficulty often encountered is in making sealable windows of the optical material – quartz, sapphire and diamond have been used. Diamond is particularly useful in anvil cells where pressures >1000 atm can be applied.

Whilst samples can be examined over a range of temperatures for reaction rates, morphological changes and degradation studies in these devices, it should be remembered that Raman spectroscopy can also be used to measure temperature by measuring the Stokes and anti-Stokes spectra and applying the Boltzmann equation (see Chapters 1 and 3).

2.5.3 Special Applications – Thin Films, Surfaces and Catalysts

Various sampling techniques have been applied to obtain spectra at micron or nanometre scales. These were reviewed by Louden [19] for visible spectrometers and include interference enhancement, surface enhancement and attenuated total reflection/total internal reflection. Figure 2.13 shows an arrangement for examining thin films using internal reflections along the film enhancing the signal. The laser needs to be launched at the correct angle to achieve internal reflection along the sample and the collection optics are set to collect a long thin image, which is effectively focused through a slit into the monochromator and hence produce a sharp image on the CCD. In this way, instead of a single-point focus, multiple excitation points are collected improving sensitivity and averaging out any imperfections in the sample.

An alternative way to achieve a similar effect is to launch the laser through a glass block with the film deposited on the top. The radiation from the wave deflected back into the glass at the film glass interface generates a field which crosses the surface (an evanescent wave), which can produce Raman scattering from the film.

Very low concentrations of material can be detected by normal Raman scattering using the evanescent field created when the excitation beam is directed close to a surface. In one simple arrangement called grazing incidence, the beam is directed along the surface and the scattered light is collected with a barrel lens, which can be imaged onto the slit of a monochromator. Another way is to use a prism setup so

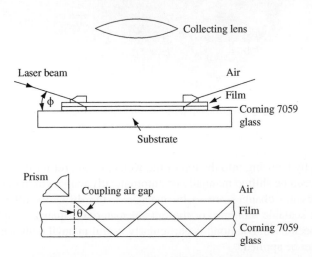

Figure 2.13. Arrangements for measuring thin films. Source: Reproduced with permission from Rabolt, J.F., Santo, R. and Swalen, J.D. (1980). *Applied Spectroscopy* **34**: 517 [20].

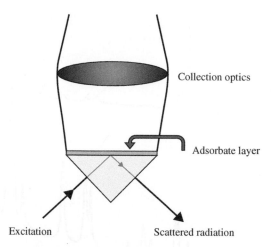

Collection optics

Adsorbate layer

Excitation   Scattered radiation

Figure 2.14. Raman scattering collected using evanescent field excitation.

that the sample to be analysed is presented as an adsorbate layer on top of one surface and the exciting beam is directed at an angle to it from below. If the angle of the laser beam is such that it is reflected from the surface, the evanescent field created at the surface causes an electric field on the area directly above the surface where the adsorbate is placed. This will cause Raman scattering from the adsorbate, which is collected on the side of the prism surface away from the exciting radiation as shown in Figure 2.14. Since the beam is reflected and does not contact the adsorbate directly, higher laser powers can be used before any sample damage occurs and there is less interference from non-Raman scattered light. To be effective, the incident radiation should cover a substantial part of the prism surface. Using a quartz crystal in a manner analogous to that used for ATR in infrared scattering with the sample mounted on top is an effective way of doing this. The direct interrogation of mono-layers on the surface by this type of approach is difficult and long accumulations are needed [21–24]. It can work well if there is a very strong scatterer on the surface or if the scattering is enhanced by resonance or surface enhancement.

Good results have been obtained using electrochemical cells when sufficient product is obtained in solution. In addition, these cells have been particularly used for surface enhanced Raman scattering (Chapter 5) where the enhancement makes it easier to observe layers on the electrode surface. Optically transparent thin layer electrodes (OTTLE), which consist of a metal grid or a transparent conducting film as an electrode in an electrochemical cell, allow the solution products to be investigated by electronic and Raman spectroscopy in the same cell at various potentials by exciting and collecting the light through the electrode. The molecular specificity of the Raman spectrum makes it possible to identify intermediates *in situ*. An example of the Raman spectra produced in an OTTLE cell when the potential is changed is shown in Figure 2.15 for a charge transfer material used in the production

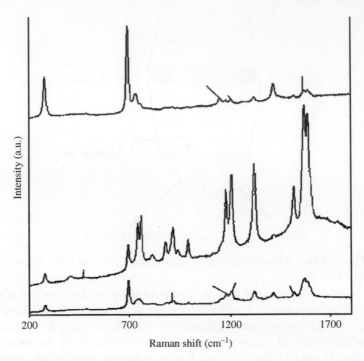

Intensity (a.u.)

200   700   1200   1700

Raman shift (cm^{-1})

Figure 2.15. Raman spectra taken from an OTTLE cell containing a solution of a charge transfer agent (top) and the mono- (middle) and di-cation (foot). The mono- and di-cation spectra are present at low concentrations but are coloured so that resonant or preresonant enhancement increases their intensity. Source: Reproduced with permission from Littleford, R., Paterson, M.A.J., Low, P.J., et al. (2004). *Phys. Chem., Chem. Phys.* **6**: 3257–3263 [25].

of light-emitting diodes. The spectra are from radicals and are very strong due to resonance (see Chapter 4).

2.5.4 Reaction Cells, Flow Through Cells, Sample Changers and Automated Mounts

It follows from the above that the ease with which scattering can be collected makes it simple to obtain scattering from solutions in glass chemical apparatus. The wide variation in scattering efficiency between analytes means that sensitivity will vary depending on the reaction studied, but without resonant enhancement relatively high concentrations are required. Further, the quality of the glassware and the curvature means that the collection efficiency can be reduced so, to study reactions, cells designed for the purpose with an optical window should be used where possible. Figure 2.16 shows such a cell for electrochemical studies. Flow cells have the big advantage that the sample is continuously refreshed under the beam minimizing sample damage. A small square tube makes focusing easier. Combining microfluidics and Raman

Figure 2.16. Electrochemical cell. Source: Image courtesy of Renishaw plc.

microscopes has led to a significant body of work using flowing solutions and this is illustrated later in Chapter 6. Automated sample changers are available for semi-continuous examination of pharmaceutical tablets (Figure 2.17), a combined macro/micro sampling stage has been developed. Microtiter plate readers are available as accessories and stand-alone instruments.

2.6 FIBRE-OPTIC COUPLING AND WAVE GUIDES

In addition to mounting samples in a spectrometer or pointing the beam directly at an object, the versatility of the technique can be extended by the use of fibre-optic probes [26]. In many applications, the use of fibre optics to separate the sampling head from the spectrometer can be a big advantage. For example, Raman spectroscopy can be used for online analysis on a chemical plant where access can be difficult and the environment not suitable for spectroscopy. Perhaps the plant is open to the elements, there is dust from delivery lorries or there simply is no space. However, with fibre optics, only the head used needs be exposed with the spectrometer housed elsewhere. Long fibres can be used to examine reactions in vessels from tens to

(a)

(b)

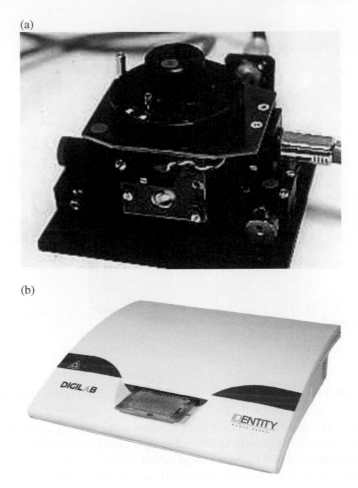

Figure 2.17. (a) Pharmaceutical tablet autochanger. Source: With permission of Ventacon Ltd. (b) Digilab identity Raman plate reader. Source: Image courtesy of Digilab Inc.

hundreds of metres from the spectrometer on an industrial plant. Further, with portable equipment it can be convenient to have a spectrometer which can be held in one hand and a simple small probe, which can be placed where sampling is required. These probes can be armoured to prevent damage. Thus, the use of fibre optics has extended the utility of Raman spectroscopy considerably.

The distance is limited by the excitation used and the quality of the glass. Silica fibres often have some impurities including iron which at some frequencies cause absorption of the excitation and, in particular, the scattered radiation and water overtones can be a problem in the NIR. In addition, while passing down the fibre optic, the intense exciting radiation will induce Raman scattering and fluorescence from

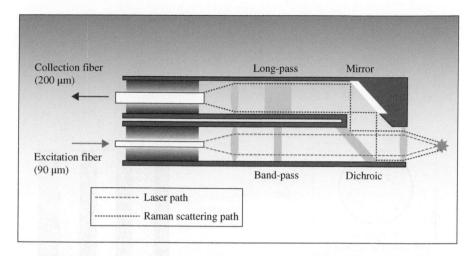

Figure 2.18. Fibre optic probe head design. The exciting radiation passes down one fibre, through a band pass filter to remove as much frequency shifted light as possible before passing through a dichroic filter. A lens then focusses the beam onto the sample. The collected light hits the dichroic filter at a different angle and is reflected through a long pass filter consisting of a number of elements to efficiently reject non-frequency shifted light and allow the Raman scattered light to be collected into the collection fibre.

the fibre-optic material. This can cause interference, particularly if the scattering from the sample is weak and a large length of cable is used. Fibre quality is therefore critical and the operator should make sure that no spurious bands are present in the spectrum. One possible arrangement of a probe head to minimize these problems is shown in Figure 2.18. To increase collection efficiency, a multimode cable in which the laser is launched down one fibre and the scattering collected through several fibres can be used.

Spectra of aspirin tablets have been reported at 50 m using band pass filters [27]. Materials which cannot be introduced into the spectrometer due to their physical size or hazardous nature can be analysed using this type of probe. It should be borne in mind that these probes like all Raman spectrometers are set to filter out non-shifted radiation but not radiation from other wavelengths such as daylight and too much light such as direct sunlight can flood the detector.

Whilst the advantages of a small sample volume have been discussed, sometimes a large sample volume is required. For example, in process analysis measurements sample homogeneity may be a problem if only a microscopic sample is interrogated, Here, a different type of probe such as the PhAT probe from Kaiser Inc. is required. In this the laser is launched through a bundle of fibres spaced out to cover an area typically 3 or 6 mm squared and with collection fibres spaced to match this. The principle is illustrated in Figure 2.19.

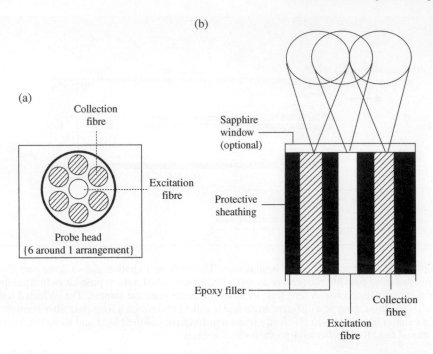

Figure 2.19. (a, b) Fibre optic probe end. Source: Reproduced from Slater, J.B., Tedesco, J.M., Fairchild, R.C., and Lewis, I.R (2001). Raman spectrometry and its adaptation to the industrial environment. In: *Handbook of Raman Spectroscopy, Ch. 3* (ed. I.R. Lewis and H.G.M. Edwards), 41–144. New York: Marcel Dekker [28].

Very small sample areas can be examined by heating, drawing out and cleaving a fibre so that an aperture typically between 100 and 50 nm in diameter is created at the tip. The light as it passes through the fibre is compressed so that it emerges and rapidly expands from the small aperture. Thus, if the tip is placed almost on a surface by an atomic force microscope (AFM) head, the effective excitation area is very small and below the diffraction limit. This process can be reversed where the laser excites the surface externally and the fibre picks up the scattered light through the small aperture. The main problem with the method is that it is inefficient and requires good Raman scatterers to be effective. This is a scanning near-field optical microscopy (SNOM).

Another way of obtaining Raman scattering is to use wave guiding. In this approach narrow tubes made with high-refractive-index tube wall materials are filled with the analyte and the laser beam is launched down the tube. Reflection from the walls keeps the beam in the tube so that it is contained in the analyte solution right down the length of the tube. The signal is then collected at the other end, passed through a notch filter and analysed in a standard Raman spectrometer. The advantage of this arrangement is that there is a very long path length and the laser irradiates the whole sample so that quite dilute solutions can be analysed. The prime requirement is that the sample has a higher refractive index than the sample tube to constrain the illuminating light within the tube and to achieve total internal reflection. Spectra of benzene and weak solutions of sodium carbonate and β-carotene have been recorded [29, 30]. One option is to use silver to produce an SERS-active surface. Detection limits of $<10^{-9}$ mol l^{-1} in low refractive index liquids have been reported [31] by using this technique.

2.7 MICROSCOPY

2.7.1 Raman Microscopes

Many modern Raman spectrometers are coupled to a microscope. This is usually done by directing the exciting beam into the top of the microscope where a mirror or beam splitter is used to direct the beam through the microscope optics and to direct the scattered radiation back into the spectrometer. The arrangement previously shown in Figure 2.1 works well for this since the exciting radiation and the scattering are already colinear. The advantage of a beam splitter is that a proportion of the light scattered from the surface can be directed towards the eyepiece enabling the position of the beam to be determined. However, this requires to be set up safely so that no bright light can reach the eye in any circumstances. Remember that it is possible that this returning radiation could include reflected light, which is still coherent and as such will be particularly dangerous. It is best to use either a camera in place of the eye piece or a commercial instrument for which the safety issues have been well thought out.

Raman microscopy provides a very powerful tool for substance analysis combining the precise sample location powers of a quality microscope with the power of Raman spectroscopy to identify compounds *in situ* without separation. Modern systems can fit on a standard bench top and offer the obvious advantages of precise *x*, *y* and *z* sample control and high-quality optics delivering a tightly focused beam to provide a high-power density from a relatively low-power laser. For solids, this also means that Raman scattering can be a highly sensitive technique with one microscopic crystal providing strong scattering. For example, a single crystal of explosive on a complete finger print has been detected. Less obvious is the fact that the very small sampling volume can be efficiently imaged into small monochromators enabling reductions in the effective size and weight of the system. Many of these systems are equipped with libraries and powerful software for mapping and imaging in two or three dimensions with colour and false colour displays similar to those used in fluorescence. The limit of resolution is usually given as half the wavelength of light and this means that the further to the red the excitation frequency the lower the resolution. Greater resolution and 3D maps can be obtained. Some examples of the use of modern systems are given in Chapter 7.

Microscopes can be set up confocally. In the arrangement shown in Figure 2.20, the microscope has a pinhole in its focal plane, which enables scattering to be collected efficiently only from light focused at particular depth, the middle of the sample volume in a solution or the surface on which the beam is focused in a solid. The pinhole filter (confocal aperture in Figure 2.20b) reduces scattering from radiation collected at other depths since it is not focused sharply in the plane of the pinhole and much of it hits the area surrounding the pinhole (Figure 2.20b). An alternative system is adopted on some instruments. In these, a slit is placed in the focal plane of the microscope at right angles to the slit of the spectrometer. In this way, the two slits although well separated in the instrument are crossed to create essentially a pinhole. In either case the intention is to discriminate against light, which may arise from depths other than the plane in which the spot is sharply focused. However, many commercial instruments use infinity-corrected objectives and so produce collimated rays from a point source, with an additional intermediate lens focusing the light onto the confocal aperture.

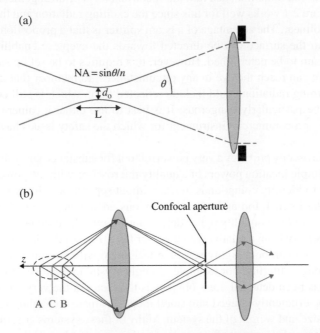

Figure 2.20. Schematic illustrating principles of confocal Raman microscopy. (a) The laser beam is focused to a diffraction-limited focal volume in the sample using a high numerical aperture (NA) objective. (b) A confocal aperture placed at a back focal plane attenuates out of focus signals by blocking rays from A and B whilst rays from C pass freely through the aperture to the detector. Source: Reproduced from Everall, N.J. (2010). *Analyst* **135**: 2512-2 [32], with permission of the Royal Society of Chemistry.

2.7.2 Depth Profiling

With the confocal arrangement, it is possible to obtain spectra at different depths into a material thus creating a depth profile. From knowledge of the magnification of the microscope objective used, it is then possible to work out the sample volume and consequently the position in the sample from which the spectra was obtained. Although this would seem in principle relatively simple, there is a considerable problem created by refraction [33, 34] as the beam enters the other material, which, in general, will have a different dielectric constant from that of the air between the sample and the microscope. This can be decreased to some extent by using water immersion and oil immersion objectives, but, in general, considerable care must be taken when estimating the true depth into the sample. However, given this limitation it is still possible to obtain some information about how a signal changes with depth in the sample. A depth profile taken with a microscope with 632 nm excitation for the polymer polyethylene terephthalate is shown in Figure 2.21. As can be seen, despite the confocal optics enough light passes through the pinhole above and below the focal plane for an effective spectrum to be obtained 2 μm above the surface.

However, the sample is not simply a polymer block but has a thin surface coating. Indicated in the top spectrum in Figure 2.21 is a band at 812 cm^{-1}, which is due to the surface layer. As a sample becomes more complex with several layers of varying thickness, the refractive index of each layer needs to be taken into account making the ray diagram complex and the sample volume of the confocal beam less easily defined. Everall has published several accounts of this issue. Figure 2.22 shows an extreme case of focusing on the edge between two materials in a layer. The spectrum contains bands from material several microns away due to wave-guiding effects.

The above effects should not deter spectroscopists from employing this very powerful and useful technique. In interpreting the spectra, due regard must be taken into account of the heterogeneity of a sample, particularly when liquids or solids with various refractive indices are present. For example, this issue requires to be borne in mind when considering more complex environments such as the spectra from tissues and cells for which examples are given in Chapter 7.

2.7.3 Imaging and Mapping

The CCD devices widely used as detectors in Raman spectrometers are essentially similar chips to those used in digital cameras and camcorders. They are arranged in an array of pixels, each of which can be individually addressed along both the x and y directions. The difference in Raman scattering is that, since the signal is weak and, compared to a camera, longer exposures are usually used, the background noise is critically important and consequently in most instruments, the sample is cooled using two- or three-stage Peltier cooling or even liquid nitrogen cooling. Some

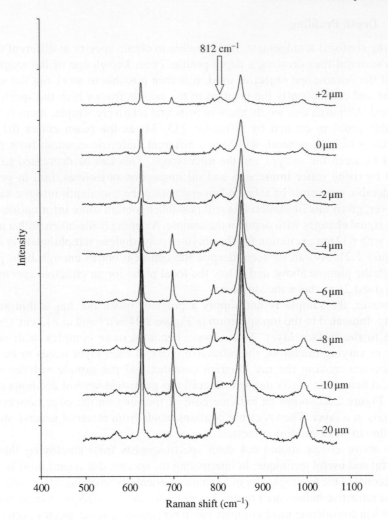

Figure 2.21. Confocal depth profile of 78 μm acrylic latex coated on 200 μm PET coating. The depth resolution was determined to be 4 μm with a ×50 objective at 632.8 nm. Spectra were collected from the coating surface (0 μm) and at 2 μm intervals through the polymer (spectra labelled −2 μm through −20 μm). The uppermost spectrum was collected when the laser beam was focused 2 μm above the coating surface, so that the Raman scattering arose from the residual, or defocused, part of the beam. Source: Reproduced with permission from Macanally, G.D., Everall, N.J., Chalmers, J.M., and Smith, W.E. (2003). *Applied Spectroscopy* **57**: 44 [35].

(a)

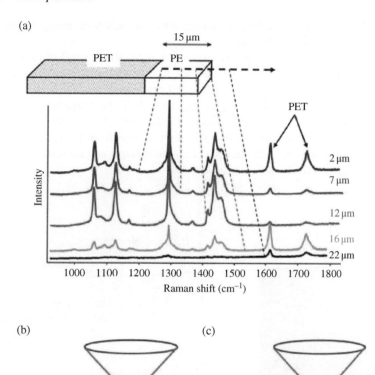

(b)   (c)

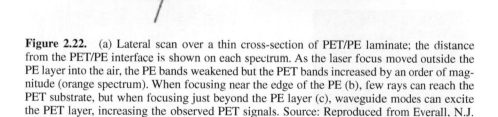

Figure 2.22. (a) Lateral scan over a thin cross-section of PET/PE laminate; the distance from the PET/PE interface is shown on each spectrum. As the laser focus moved outside the PE layer into the air, the PE bands weakened but the PET bands increased by an order of magnitude (orange spectrum). When focusing near the edge of the PE (b), few rays can reach the PET substrate, but when focusing just beyond the PE layer (c), waveguide modes can excite the PET layer, increasing the observed PET signals. Source: Reproduced from Everall, N.J. (2010). *Analyst* **135**: 2512-2 [32], with permission of the Royal Society of Chemistry.

less-expensive spectrometers employ chips which can be just a linear array either with no cooling or with single-stage cooling. In this case the scattered light from a point focus is separated into individual frequencies by the grating and focused as a line on the CCD so that each separate frequency can be detected separately. The 2D arrangement can operate similarly but also enables imaging and allows a line focus to be used with the different frequencies recorded for each point along the line in the direction perpendicular to the line.

An alternative way of collecting Raman scattering is that instead of using a mono-chromator to split up the different frequencies, a set of filters can be used, in a manner analogous to that used by Raman in the initial experiment. In this arrangement only light of a particular frequency range chosen to include the frequency of one of the major vibrations of the analyte can pass to the detector. The detector operates like a camera recording a Raman image of a specific vibration from the focused area. This is called imaging. Figure 2.23 shows a photomicrograph and corresponding Raman

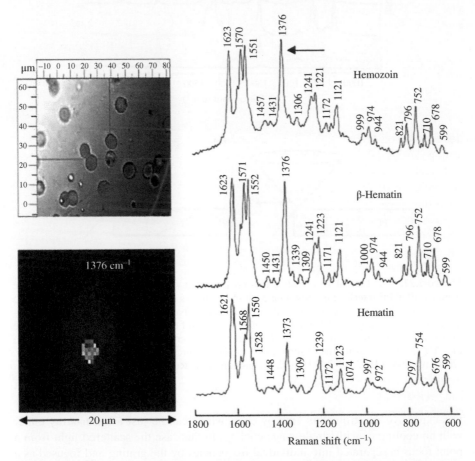

Figure 2.23. Raman imaging. Photomicrograph and corresponding Raman image of the $1374\,cm^{-1}$ band. Source: Reprinted from Wood, B.R., Langford, S.J., Cooke, B.M., et al. (2003). *FEBS Letters* **554**: 247–252 [36].

image of the 1376 cm^{-1} band from a parasite's food vacuole along with the spectra of hemozoin, and related β-hematin and hematin all acquired using 780 nm excitation [36]. The spectrum of hemozoin is similar to the spectrum of β-hematin at all applied excitation wavelengths. Resonance enhancement of A_{1g} modes, explained in Chapter 4, including ν_4 at 1374 cm^{-1} involves excitonic coupling between linked porphyrin moieties in the extended porphyrin array. The intensity of the band enables the investigation of hemozoin within its natural environment in the food vacuole.

If the sample is small enough to be imaged, imaging is quick and can be compared directly to a photograph taken through the instrument with the sample *in situ*. However, the filters only provide limited spectroscopic data. They cover broad frequency ranges relative to the natural line width of a Raman band so that a number of vibrations may contribute to the scattering and if more than one range is required, the filters have to be changed.

Modern instruments often use hyperspectral imaging, examples of which are given in Chapter 7. Briefly the line focus described above is used so that the different frequencies of Raman scattering are presented along the other axis using all of the pixels of the 2D array. Using an electronic mirror, the beam is then scanned across the surface collecting the Raman scattering at many line positions giving a full spectrum for many closely spaced points on the sample. Using the z axis this process can be repeated at different depths to provide 3D information on the sample. From the wealth of data collected, it is then possible to select out a peak and create a 2D or 3D image of the sample for that vibration. This process is called hyperspectral imaging. This a very attractive method but previous comments on refraction, photon migration depth of penetration and imaging should be, but often are not, considered. In particular, the effect of the sample on depth penetration described above should be taken into account in order to optimize the image quality and spectral integrity [37].

An alternative simpler method is to map the surface. The sample is mounted on a motorized *xyz* stage connected to the spectrometer software. It is then possible to collect spectra point by point across a predetermined area of the sample but this is time consuming. Figure 2.24 shows a typical simple map. One is a black-and-white image of the surface with each pixel shown being a point at which a spectrum was taken. The lighter the pixel, the more intense the Raman scattering recorded. The other map shows a 3D representation. These maps were obtained from a surface in which a number of small particles, which give extremely strong Raman signals, had been deposited. The position of the particles can clearly be seen from the peaks shown in the map.

In all these methods a significant amount of radiation can fall on the sample. Particularly, if any absorption is occurring this can lead to some heating and sample degradation. Depending on the nature of the sample the exposure times and laser power require to be set within a range, which prevents appreciable sample damage.

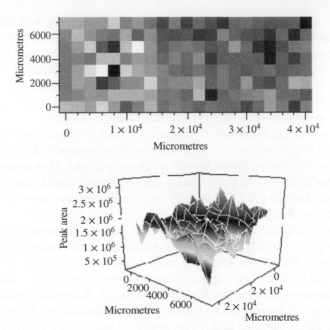

Figure 2.24. Raman maps – pixel map (top); 3D map (bottom). Source: Reproduced with permission from McCabe, A., Smith, W.E., Thomson, G., et al. (2002). *Applied Spectroscopy* **56**: 820 [38].

2.8 CALIBRATION

So far, we have considered the spectrometer components and sample presentation. Before continuing to examine and manipulate the data, a question which should be asked is, 'How do you know that your instrument is working correctly and consistently?' It is a question regularly asked of industrial spectroscopists, specifically by regulatory authorities. The questions are often and rightly asked by non-scientists. The pharmaceutical industry, in particular, has to register new products before sale to the public. The regulators wish to know that the measurements have been made correctly, on correctly working instruments which will give the 'same' answer today and tomorrow on the 'same' or similar samples. The question is particularly important if quantitative work is carried out. Apart from regulatory requirements, industrial spectroscopists often require quantitative methods to be transferred between instruments. The search for a simple standard and method of calibration proved difficult but at least one daily check had been published by McCreery [1].

Most of the checks that are carried out are to ensure that the *x*-axis or wavenumber position is correct. The phrase 'calibration' is often used but most spectroscopists do not carry out fully calibration checks which are often done only by optical engineers.

The checks that are carried out are better described as performance checks and use standards to check that the frequency and intensity are in specification. Barium sulphate has a strong band at 988 cm^{-1}, diamond a band at 1364 cm^{-1} and silicon a band at 520 cm^{-1}, which are now used by some instrument manufacturers. In addition, indene [39], cyclohexane and sulphur have well-known band positions as measured on dispersive instruments. One peak such as the peak on silicon is often used as a day-to-day check but is better to check with as sample such as cyclohexane which gives a number of peaks across the range of wavelengths to be studied.

The height of the peak or peaks of the calibrant should be and usually is measured, but calibrating relative peak heights is rarely mentioned. For dispersive instruments the effect of different frequencies on intensity has been pointed out earlier (Figure 2.4). The fact that the instrument software should effectively correct for changing instrument efficiency with frequency requires to be checked. For FT NIR Raman spectrometers the situation is worse. Whilst the sulphur spectrum maintains relative band strengths, indene bands vary greatly in relative intensity with laser power (Figure 2.25). Whilst the spectra appear very similar, the ratio of the 2890–1550 cm^{-1} bands does not change linearly with a change in laser power from 10 to 350 mw. A number of compounds with absorption bands in the NIR spectrum above the laser line at 1064 nm show this effect. This appears to particularly affect compounds with aliphatic hydrocarbon groups. The bands with a Raman shift of ~3000 cm^{-1} are actually scattered at a true frequency of 6398 cm^{-1}, which is equivalent to 1562 nm. This is very close to the broad NIR aliphatic hydrocarbon overtone bands at ~1666 nm. Halogenated dienes have been suggested as a possible standard. CH stretches, and others near to the limits of the detector range, are also frequently strongly attenuated compared to spectra excited with a visible light laser. Bands are likely to be severely attenuated

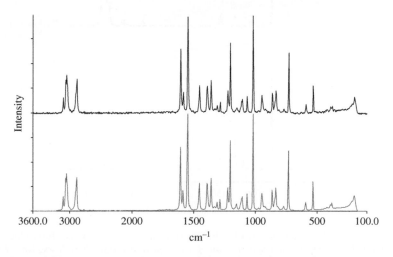

Figure 2.25. Spectrum of indene at 1064 nm–10 mw (top); 350 mw (bottom).

when they occur close to the cutoff edges of the filter used to block the elastically scattered radiation occurring at the exciting line frequency. Standards based on fluorophores have also been proposed for spectrometers with visible laser sources.

A neon lamp on the beam axis can provide a wavelength calibration standard. Halogenated dienes and cyclohexane have been suggested as possible Raman wavenumber standards [40]. The spectrum can have a varied response to intensity depending on excitation frequency, instrument and detector and may well have to be corrected as illustrated by the spectra shown in Figure 2.26. An ASTM standard (ASTM E 1840-96 Reapproved 2014) has now been established for calibrating the Raman shift axis. Eight common chemicals – 1,4-bis(2-methylstyryl)benzene, naphthalene, sulphur, 50/50 (*v/v*) toluene/acetonitrile, 4-acetamido-phenol, benzonitrile, cyclohexane and polystyrene had the Raman spectra recorded by six different laboratories using both dispersive and FT spectrometers. Apart from a few of the values at high and low frequencies, standard deviations of <1 cm^{-1} were reported.

Tungsten lamps were often quoted as the way to measure the variable instrument response with wavelength. Unfortunately, the lamp energy is dependent on temperature and this varies with the lifetime of the bulb. For an accurate calibration, the filament temperature would have to be measured. The use of tungsten lamps and glass

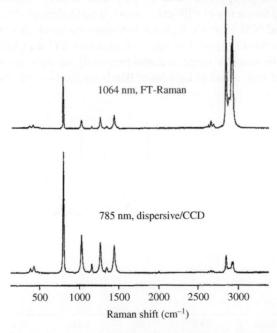

Figure 2.26. Cyclohexane uncorrected for instrument response. Source: Reproduced from McCreery, R.L. (2000). *Raman Spectroscopy for Chemical Analysis*. New York: John Wiley & Sons, Inc. [1].

filters had been proposed by NIST [41] and adopted by instrument manufacturers [42] to overcome this issue, particularly for transfer of quantitative methods. To calibrate the *y*-axis, a simple, practical calibration standard for instrument response correction is required. ASTM has published a number of standards for checking instrument performance. These include E2911-13 Relative Intensity Correction, E1683-02(2014)e1 Standard Practice for Testing the Performance of Scanning Raman Spectrometers, E2056-04(2016) Standard Practice for Qualifying Spectrometers and Spectrophotometers for Use in Multivariate Analyses, Calibrated Using Surrogate Mixtures. There is also E1654-94(2013) Standard Guide for Measuring Ionizing Radiation-Induced Spectral Changes in Optical Fibers and Cables for Use in Remote Raman Fiber Optic Spectroscopy.

Whilst standards are now more available, there is still not a universally accepted, easy to use, sample which calibrates both wavenumber and intensity in a single spectrum. The luminescent standards have to be used in the same sampling geometry as the sample of interest. The wavenumber's position can be affected by several instrument features, particularly in FT systems [43].

2.9 DATA MANIPULATION, PRESENTATION AND QUANTITATION

Having organized the Raman experiment with regard to sample presentation and instrument operation, we need to consider how the data will be generated and manipulated. The latter will depend on the use to which the data is put. As already stated, the phrase 'interpretation of Raman spectra' is used in many different ways. In qualitative analysis the spectrum of a molecule can be the subject of a full theoretical interpretation in which every band is carefully assigned or of a cursory look to see if a specific band or pattern of bands is present or of a library search by the software to produce the interpretation 'Yes that is acetone'. Alternatively, the spectra could be employed to monitor or determine composition in a quantitative way by plotting the intensity of a single band or analysing the entire spectrum with a data analysis program such as principal component analysis (PCA). Whichever way the data is used, the manipulation of the data, which has occurred during production of the spectrum, has to be considered.

2.9.1 Manipulation of Spectra for Presentation

Raman instruments are single-beam instruments that are operated in the vast majority of cases without the use of a background reference spectrum. The instrument is calibrated before delivery to compensate for changing detection efficiencies with wavelength and normally the user does not need to be concerned. However, it is important to be aware that problems can arise. As discussed earlier instrumental features such as degraded filters or beam splitters can produce spurious peaks close to the cutoff, and faulty changeovers between filters can alter intensities. In FT spectrometers, the raw data is not a spectrum but an interferogram. This is computer manipulated before presentation as a spectrum. The relative strengths of the bands in the $3000\,cm^{-1}$ region are particularly affected in FT Raman and as discussed above the use of 792 or 850 nm excitation with visible Raman systems can also affect relative intensities, particularly with absorbing or turbid samples. Another feature is the direction of polarization of the laser beam but this will be discussed in Chapter 3.

Two particularly common spurious bands regularly observed are peaks due to cosmic rays hitting the CCD in a dispersive system and side bands from the laser source which because they are at a different frequency to the main laser beam can pass through the system. Bands from both can often be recognized immediately since they are very sharp. Cosmic rays can be excluded by taking multiple accumulations and averaging. The software on most instruments will then remove peaks obtained in only one accumulation. Laser side bands should be removed by using a filter in front of the laser. However, modern Raman systems are very sensitive and the authors have seen these bands occur when the sample presented gave little to no scattering and the detector sensitivity was very high.

For precise measurements, background correction can be carried out with a white light source. In the ideal world, this would have a known, invariable temperature. In practice, this does vary with time and can cause variations in the background. These effects are most critical for quantitative measurements rather than qualitative measurements. The effects of apodization and resolution can be seen in the spectra from FT Raman instruments.

A major advantage, and also a problem, is the flexibility of the software used following data capture. It is often easy to produce an apparently strong spectrum by simply changing the intensity scale. This is often carried out automatically by the instrument. If care is not taken to read the Y scale, the information that the spectrum is weak can be missed. A weak spectrum may be due to too little sample, poor preparation, the fact that the 'sample' is loaded with a diluent such as salt or simply that the sample is a poor Raman scatterer. In the last case, the spectrum may be from an impurity due to a strong Raman scatterer in the sample matrix.

Spectra of widely differing intensities are often scaled for comparison, for example, by 'normalizing' a major peak to the same value for each spectrum. Data management systems on some instruments may automatically scale spectra so that the strongest peak stretches to the top of the screen and are presented as 'normalized'. The argument is used that sampling methods, for example, quick spectra taken from a large number of different powders, do not enable intensity comparisons between spectra or that spectra with widely differing intensities cannot be easily presented for comparison in one diagram. Depending on the information required this may be legitimate, but absolute intensity data is lost and comparisons tend to start with the assumption that the major peak in each sample is of comparable intensity. In many samples, large differences are due to significant differences in Raman cross-section or in some cases to poor sample preparation. Preparation of a new sample can often check this. In any event the spectroscopist is better to have the true spectra and work to obtain the correct interpretation from there. The experienced spectroscopist will look at the noise present on the spectra away from the main peaks to judge the relative intensity of the peaks in the spectra in their original form. However, smoothing routines, which on many occasions can be quite useful, will remove most noise from the spectra often preventing it being used to estimate intensity changes between spectra.

A better way to scale spectra for comparison purposes is normalizing against a standard. Ideally, the standard is mixed into the sample either by dispersing a powder in a powder sample or by adding a solvent to a liquid. In both cases, it is essential that this does not change the chemistry, that the calibrant is properly mixed and is sufficiently intense to be identified easily in the spectrum. Despite the fact that most instruments are single beam they should be stable. By running a standard such as cyclohexane in a cuvette a number of times will prove stability and then run at least one spectrum before and after sample measurement. Of course, it is always possible to set up a double beam arrangement or use a cell alternating the sample and standard by spinning it but this is usually too time consuming for most studies.

Baseline subtraction is commonly used to help present Raman spectra and very good routines are available. However, the baseline can contain important information. For example, with coloured samples, the spectra are often presented as having a flat baseline but in the original spectrum, depending on the conditions, broad peaks may be present due to fluorescence. These bands can be genuine and give useful data which is removed by the baseline correction. Further, in both coloured and colourless samples sloped baselines, particularly if they vary from sample to sample, may suggest a problem in the sampling such as misaligned cells, presence of particles, absorption of scattering at certain wavelengths, etc. Thus, these routines should be used with an understanding of what may be occurring!

If quantitative work is being carried out, another way of correcting for a background slope is to carry out derivative spectroscopy. As can be seen in Figure 2.27 the spectrum of the second derivative has a flat baseline but identifying individual bands can be more complex.

Other routines can be used to deconvolute broad spectra with features as can be seen in Figure 2.28. Good judgement has to be made as to when to stop if the number of components is unknown otherwise spurious bands may be added to the spectrum.

Further, smoothing routines can be used to reduce noise and clarify weak bands in a noisy spectrum. However, this must be used with caution. Overuse can lose shoulders, remove some bands completely and provide strange-shaped bands. Figure 2.29 shows a spectrum with several degrees of smoothing. In addition, very small often noisy features of the spectrum can be presented as larger than they are by selecting out, for presentation, a small part of the spectral range and smoothing out any noise present. This will produce a smooth-looking band, which, since no strong band is present in this region gives the impression that it is a major feature. This may be a

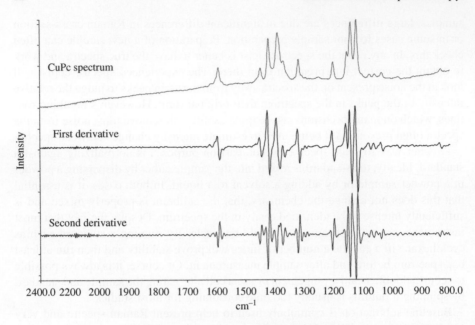

Figure 2.27. Derivative Raman spectra.

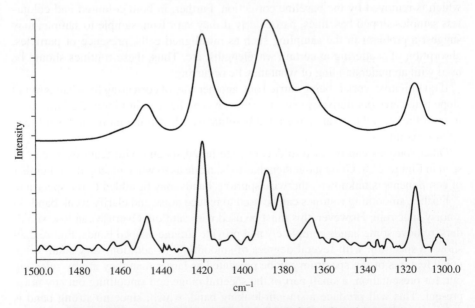

Figure 2.28. Deconvolution of Raman bands from a broad spectrum.

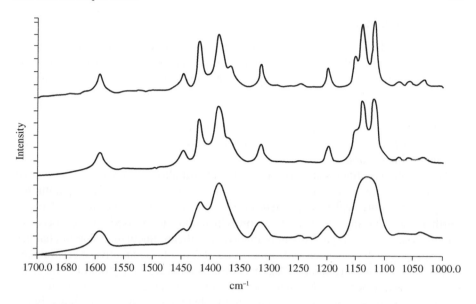

Figure 2.29. Smoothed Raman spectra with the foot spectrum showing the danger of oversmoothing.

correct approach and the band may arise from a weak scattering major species in the sample but it is also possible that it arises from an impurity or other spurious cause.

2.9.2 Presentation of Spectra

Spectral presentation is generally not an issue. Raman spectra are usually presented as just the Stokes spectra with the anti-Stokes spectra omitted. The only inconsistent feature is in the way in which the wavelength scale is displayed, sometimes from high to low wavenumber but often from low to high wavenumber. There are semantic debates as to which is correct. Purists say that all graphical scales should be displayed with the lowest absolute energy value at the origin. Others say that Raman scattering is a shift and not an absolute measurement and it is the absolute energy of the vibration which should be plotted with the lowest vibrational energy closest to the origin. For comparison with infrared spectra, this format is preferred since infrared and Raman spectra from the same sample can be overlaid and band positions compared. The relative intensities of Raman bands should be independent of the instrument but the absolute intensity varies from instrument to instrument. Calibration as described above is essential for quantitative measurements to enable comparison with data from other instruments. It is common for qualitative comparisons just to display the intensities using, for example, counts per second from the CCD. Particularly when data has been manipulated, the scale may be described as in arbitrary units and in this case it may result in different unspecified scales for each spectrum.

2.9.3   Quantitation

The data-handling procedures have largely concentrated on the effects on the qualitative aspects of the spectrum, but Raman spectroscopy is also used quantitatively. For most uses of Raman scattering, the essential feature is the ability to detect the spectrum. The frequency values are given with reasonable accuracy. Intensities, on the other hand, are usually treated as relative intensities or described as 'strong', 'medium' and 'weak'. This is sufficient information if what is required is to use the spectrum as a fingerprint for the molecule, or molecules, from which the scattering occurs.

Quantitation for liquids is best carried out with a quality instrument which is stable. Raman spectrometers are single-beam instruments and consequently any quantitative analysis procedure will be dependent on the stability of the laser and the detector. Further, the same instrumental condition will require to be used on each occasion the analysis is carried out. This may involve significant fluctuation due to changes, for example, in laser power with time. As a result, all quantitative measurements using Raman scattering should make use of a calibrant, which should be run at the same time and, where possible, interspersed with the samples used in the quantitative procedure. Many systems use a silicon calibrant but the main peak is at relatively low energy at about $550 \, cm^{-1}$ and this is often used as the calibrant in the analysis. It is usually better to use a second calibrant such as cyclohexane so that there are multiple bands some at higher and some at lower wavenumber than the analyte band to be measured. In principle, the instrument should cope with the difference in frequency, but the authors have observed changes caused on one occasion by a slipping stage on a grating in the instrument and in another with the use of NIR radiation where the detection of the Raman lines was close to the edge of the range for the detector.

One problem is that the flexibility of Raman sampling can mean that reproducibly replacing a sample may be difficult. The use of focused beams in Raman scattering to obtain higher power means that relatively small volumes of solution are usually interrogated. In clear or lightly coloured solutions, the most common approach is to use a 1 cm cuvette in a fixed holder and with a focused beam position. The volume of the sample within the cuvette actually sampled is usually of the order of microlitres. To obtain effective Raman scattering from this volume, the beam must first of all pass through the cuvette and therefore be refracted by it and then through the media. The scattered radiation then repeats the process on the way back to the detector. The depth at which the sample is focused can alter the signal, and any misalignment of the cell which causes a slight displacement of the laser beam can also affect signal intensity. Thus, it is essential that a stable holder is built which defines the position of the sample in relation to the collection optics. It is also essential that the instrument parameters be set in exactly the same position on each occasion.

Sometimes a probe head is used to focus directly onto a flowing line or into a reaction vessel. To do this effectively the probe and vessel must be fixed in position and the flow in the line or vessel controlled. The glass wall of the line or vessel is likely to be curved increasing diffraction problems and since it is not an optical surface it can cause partial beam defocusing. How important this is depends on the acceptable tolerance. Measurement of whisky directly through bottle requires very precise measurement. A major factor is the thickness and evenness of the glass in each bottle and the precise orientation of the bottle to the probe. Turbid solutions and particles in suspension make quantitation more difficult since in solution optical transparency will vary with time.

In solids, quantitation is usually applied to powders and many variables require to be taken into account. Particle size and density are obvious and photon migration will quickly disperse the beam so that only a narrow surface layer is analysed. If possible, transmission scattering as outlined previously may be the better technique since a larger amount of the sample is interrogated and the approximate sample volume calculated. For backscattering a standard is usually thoroughly dispersed into the sample and scattering collected from as much of the sample as possible. Several measurements may be taken across the sample to ensure a representative result or the sample can be mapped as described before and the results averaged. As with other optical techniques, it is easier to quantify solutions or gases rather than solids where the nature of the solid can have a large effect on the spectra obtained. The use of the microsampler attached to a microscope, or the use of a system without a microscope, so that a representatively large amount of the solution is detected is also a help.

Clearly, a double-beam approach would be more effective for quantitation. One possible approach is to use a cell split into two with one half filled with sample and the other with the standard. The cell is spun so that the standard and sample spectra are recorded regularly over a period of time. The result is then obtained from the average accumulated signal. However, these can be difficult to fill and to use and modern machines are sufficiently reliable for single-beam use with a calibrant in most cases. One important variable to check is the laser power which if not automatically compensated can drift over a day. Recently, instruments which can record quantitative spectra in the standard 96 or 384 well-microtitre plates widely used in biology have become available.

Whichever way the quantitative measurements are carried out, it is essential that all the spectrometer conditions, sample concentrations, sampling volume and, where relevant, cell window material or reflection angles must be noted, if the measurement is to be repeated or transferred to different instruments. Relative band intensities will vary with different wavelength laser sources because of the fourth power law. As already mentioned self-absorption can affect band intensities, particularly in NIR Raman. Relative intensities can also be enhanced by resonance as described in Chapters 4 and 5. Temperature may also be critical in some measurements, especially at high laser power or with a strongly absorbing sample, which could undergo thermal degradation.

Having obtained the spectra or set of spectra, band intensities may be calculated in various ways. The most common method is to measure peak height but some instruments can calculate band areas. Since there is often some background not due to Raman scattering, a baseline must be established under each peak which, ideally should be low and level, and take account of band shape, neighbouring overlapping bands and possible fluorescence. The absolute position of the baseline for each peak is not critical but the method of determining it must be consistent in both calibration and test samples. Many instruments will do this automatically. In more complex situations such as quantitation in a biological matrix, the background can be significant and sloped making the exact position of the baseline difficult to determine exactly and in this case care must be taken with automatic programs which could change reference points as the analyte concentration increases.

Relative strength determinations can be made, in multicomponent samples, by measuring the ratio of intensities of relative bands. However, the relative Raman scattering cross-section should be borne in mind for each component. The calibration curves should ideally be constructed from similar samples of known composition. For more than two components, synthetic mixtures do not usually provide adequate calibration curves. Real samples can contain minor components which contribute to band shape and size. Components may also interact with each other and solvents to produce variations in peak position, shape and size. As already mentioned, particle size, self-absorption and depolarization ratios can all affect relative band strengths.

So far, measurements have been described from a spectroscopist viewpoint, but an analyst is more likely to use programs which collect all the data from a sample and use it all to get the best result. For example, the sharp fingerprint nature of Raman scattering makes it easy to discriminate different species in a mixture but where several are present this can become more difficult by eye although the information to do this is present. In this case PCA can be very effective and is widely used.

There are a large number of quantitative software packages available. Some can be used for composition analyses which attempt a simple least squares fit, through principal component regression (PCR) to partial least squares (PLS) modelling. Besides these, spectral enhancement and band resolution packages are available on many instruments. Simple derivative spectroscopy has already been mentioned, but Fourier domain processing and curve fitting routines, sometimes in complex combinations, can also be employed. All must be employed with an understanding of the applicability of the package used to the problem being studied. Otherwise, the result can be biased towards the operator's preference.

A number of texts delve deeply into the mathematics of the quantitative aspects of vibrational spectroscopy. We have only highlighted here the features which are of particular note for Raman spectroscopists.

2.10   AN APPROACH TO QUALITATIVE INTERPRETATION

In Chapter 1, the basic theory and approach to interpreting a Raman spectrum were set out. In this chapter, we have considered various instrument features such as the source wavelength, accessories which give effective sampling and how data production can affect the final spectrum. All these factors should be borne in mind before any attempt is made to interpret a spectrum. In fact, some of these parameters may have been specifically chosen to enhance a particular feature of interest. In Raman spectroscopy, whole techniques can be devoted to specific enhancement as will be seen with the SERRS effect in Chapter 5. If the basic structure of a molecule is known, then the theory expounded in Chapters 1, 3 and 4 will assist the spectroscopist to make great progress towards gaining chemical, physical and even electronic information about the state of the molecule. Subtle and not so subtle changes in bands can yield extensive specific information about the molecule. Vibrational spectroscopy is often used in attempts to identify unknown materials, to characterize reaction by-products and to follow reactions. Although Raman spectra are often simpler and clearer than infrared spectra, they can be less easy to fingerprint since some groups do not give strong bands and there are far fewer published, recorded reference spectra for direct comparison with unknowns. However, like most tools when used with skill and the correct approach, Raman spectroscopy can be of great assistance in identifying unknown materials or components. To do this successfully the maximum information about the sample should be obtained and borne in mind during the analysis and it is essential to be aware of problems which can lead to an erroneous result.

2.10.1 Factors to Consider in the Interpretation of a Raman Spectrum of an Unknown Sample

In a practical interpretation, it is essential that all available information is used and that the possibility of contamination is considered. There are a number of examples in the literature of this simple precaution being ignored and important conclusions drawn on data which subsequently were shown to have arisen from a contaminant. Whilst in both Raman and infrared spectroscopy, interpretation of the spectrum requires knowledge of all the factors, which may be affecting the spectrum, Raman spectroscopy has fewer complexities. Sample preparation is often zero with samples being examined as neat solids, liquids or gases, with only a few possible artefacts. As stated above, some instrumental effects such as cosmic rays and emission from room lights and in particular strip lights and cathode ray tubes can show up in the spectrum. These appear abnormally strong in weak spectra where scale expansion has been used. Some but not all of these features can be recognized because the bandwidth is narrow and if a number of accumulations are averaged, the instrument software should remove cosmic rays, but it is essential that thorough checks are made for the presence of these peaks.

It is important not to lose sight of the overall picture. If simple information on the nature of the sample is ignored, answers can be generated which common sense tells us are impossible. The very different intensities of Raman scattering from various vibrations from different molecules in a matrix can easily lead to this sort of wrong interpretation. A polymer bottle may contain sulphur. Polymers are weak scatterers whereas sulphur is a strong scatterer. The fact that the spectrum is dominated by the sulphur peak does not mean that the polymer is largely sulphur. This is a rather trivial example but this mistake is easy to make when two organic molecules are present in a matrix.

Raman spectra are not obviously dependent on the chemical and physical environment of the sample being examined. Whether the molecules are in a gaseous, liquid, solid or polymeric form is not easily apparent from the spectrum, but the physical state does affect the overall strength and band shape. In general, crystalline solids give sharp, strong spectra whilst liquids and vapours tend to have much weaker spectra. Pressure, orientation, crystal size, perfection and polymorphism may affect the spectra, but the changes can be subtle. Raman spectra are, however, particularly temperature-sensitive. Broad bands in Raman spectra tend to be due to fluorescence, burning, low resolution or weak bands that have been enhanced, e.g. from glass or water. Chemical groups may also respond to hydrogen bonding and pH changes but these changes tend to be shown in peak shifts rather than changes in band shape.

So having recorded the spectrum, we need to develop an approach which will help as much as possible towards solving the problem. By sequentially going through the next steps the chances of making an error in interpretation will be much reduced but success is not guaranteed!

2.10.1.1 KNOWLEDGE OF THE SAMPLE AND SAMPLE PREPARATION EFFECTS

A lot of information can be gained by understanding the way in which a sample arrived in the state presented for examination. The analyst should consider the following questions. How was the sample produced? What is known of the reaction scheme? Are there possible side reactions? Could solvents be present? Did work-up conditions introduce impurities? What was the type of equipment the sample came from? (Grease, drum linings, coupling tubes and filter aids can all appear in or as the sample spectrum.)

Solids – Is it 'dry', or a paste? Has it been washed with a solvent or recrystallized?
Liquids – Are they volatile, are they alkaline, neutral or acidic?
Vapours – What temperatures/pressures are involved?
How pure is the sample thought to be? Is any elemental information available, does the sample contain N, S, or halogens? Could these come from an impurity?
Are there likely polarization, orientation or temperature effects?

The answers to these questions are not always available but these points should be kept in mind if the spectrum does not appear as expected. The spectrum of a sample without any known history or source should be approached with great care. Something is always known even if it is only physical form and colour.

Handling the sample may affect the resultant spectrum; as mentioned in previous sections, the information required may dictate the sample preparation and/or presentation method. Knowing the method should provide some information about a sample but beware.

- Solids – Is it neat, a halide disk or mull? If it is a mull, mark off any bands from the mulling agent. Is the sample a neat powder, could this produce orientation or particle size effects? If the sample has been diluted because it has a strong colour why say it is a colourless material?
- Is the sample is in a container? Mark off the bands due to the vessel walls, e.g. glass, polythene.
- In cast or polymer films, is there any solvent trapped or encapsulated? Can the polymer film have orientation?
- Liquids – Is the sample a pure liquid or a solution? If the latter, mark the solvent bands.
- Microscopy – Are the bands real, or due to the mounting window, e.g. diamond?

2.10.1.2   INSTRUMENT AND SOFTWARE EFFECTS

The above-mentioned approach checks that all the bands and overall shape of the spectrum are not affected by the samples and the method of preparation; however, extra bands and anomalies may occur from instrument or software artefacts.

- Which laser line is used as a source, are resonance or self-absorption likely to affect band strengths?
- Does the spectrum really have a flat background or has a software background correction removed fluorescence, and destroyed information?
- Is the spectrum as strong as it appears? Check the scale and check for expansion routines.
- Has a smooth function been applied which leads to loss of bands normally resolved?
- Modern data systems display and plot information on data manipulation. Has this been applied? The lack of printed information does not mean manipulation has not occurred.
- Are the broad bands in the Raman spectrum due to fluorescence or burning?
- Are these sharp bands in the Raman spectrum which could come from cosmic rays or neon room lights?

2.10.1.3   THE SPECTRUM

Once all the information on the sample history is acquired, and all possible distortions and artefacts have been identified, or dismissed, interpretation of the band positions and strengths should begin.

- Look at the total spectrum as a picture, does it look as expected from the sample. Are the bands broad or sharp? Are they strong or weak? Is the background sloping or flat? If it appears correct continue with band position interpretation.

- Start at the high wavenumber end, in the 3600–3100 cm^{-1} region; are there any —OH or —NH bands? Refer to Tables 1.1–1.5 to determine the type, and for confirmation, look for related bands in other parts of the spectrum, e.g. amides have carbonyl bands as well as —NH bands. These bands can be weak in Raman spectra and are easily missed or not seen.
- In the 3200–2700 cm^{-1} region, are there unsaturation or aliphatic bands present? Unsaturation is usually above 3000 cm^{-1}, aliphatics below. If aliphatic bands are present, are they largely methyl or longer —CH$_2$— groups? Again refer to the tables for confirmation by other bands.
- Are there bands in the cumulative bond (e.g. —N=C=N) region 2700–2000 cm^{-1}?
- Are there bands in the double bond (e.g. —C=O, —C=C—) region 1800–1600 cm^{-1}? In the Raman spectrum, unsaturated double bond bands are generally stronger and sharper than carbonyl bands. Infrared active bands can also appear in this region.
- By these checks, it should be established if the spectrum contains aliphatic, unsaturated or aromatic groups. Multiple bond bands or carbonyl bands should also have been identified. Look at the rest of the spectrum for strong bands. Do they correspond to bands in the tables?
- The region below 1600 cm^{-1} contains many bands largely due to the fingerprint of the molecule. Structural information can be gained from this region, but bands are mainly due to the backbone of the molecule. Selected phenyl ring modes and groups such as the azo group can be identified. Other groups with bands in this region tend to be oxygenated organics, e.g. nitro, sulpho or heavily halogenated hydrocarbons. Inorganics have sharp Raman bands in this region (see Appendix A).
- Besides information identifying groups, is there negative information from bands that are not present? If the 3200–2700 cm^{-1} region contains only very weak or no bands, then this negative information could be due to unusual species such as the halogenated species mentioned, that the Raman bands of these groups are too weak or that the sample is inorganic.
- Having established the possible groups present in the spectrum, can they be combined into a molecule which can be expected from the known chemistry and/ or from the knowledge of possible impurities?

Always, wherever possible, crosscheck the interpretation by visibly matching to a reference spectrum of the molecule or of a very similar structure. Never trust peak list or computer search printouts without visually matching the spectra. Finally check again if the answer makes sense with the sample. Is a red powder really ethanol? If this general procedure is followed, then the maximum information will be obtained from the examination, and errors will be minimized.

2.10.2 Computer-Aided Spectrum Interpretation

If the objective of recording the Raman spectrum is to identify a compound the simple way to do this is to use a reference library. Most Raman instrument manufacturers now offer their own library search routines which contain prerecorded reference spectra and which can be expanded with the user's own recorded spectra. The major library publishing companies also offer their own versions of electronic libraries as stand-alone library search systems and there are some Internet searchable libraries. However, when using these systems, the results should be visibly crosschecked by comparing the spectra rather than accepting a computer listing. It is always better to check the assignment as described at the end of Section 2.10.1.

For more detailed interpretation, calculation of the spectrum, most commonly using the density functional theory (DFT) approach, can be carried out. These calculations can give good approximations to the real spectrum in both energy and intensity if carried our correctly with a large, correctly chosen basis set of functions. The results need checked and the authors have also been faced with choosing between several attempts using different basis sets. There are some simple steps which can be taken to aid evaluation of the calculated result.

1. Recognize that these calculations are usually for the isolated molecule with no environment (often called gas phase). The calculation can be improved by adding a solvent layer but it will still not be the same as the environment in the experimental sample. Therefore, some differences may be expected and the calculation should not be manipulated to try to remove them.
2. There are exceptions but most Raman spectra except resonance Raman spectra have no overtones so the region from between 1600 and 1700 cm^{-1} to just below 3000 cm^{-1} should be clear except for specific bands such as CC and CN triple bond stretches discussed in Chapter 1. For a specific molecule, inspecting the structure to see whether or not both aliphatic and aromatic C—H stretches are present enables the upper limit to be more accurately defined. A similar approach can be used to define the lower limit by assigning the highest band expected in the 1600–1700 cm^{-1} region. Does the calculation fit the gap correctly to within a few percent? If it does, this gives some confidence that the results can be used.
3. Are each of the main peaks at approximately the correct energy and due to the type of vibration which would be expected from initial assignments using the approach in the previous section? If this is not the case it does not mean that the calculation is incorrect, the problem could be with the calculation, the initial assignment or the difference in environment pointed out in (1) above.
4. If possible, both calculate and obtain experimentally the infrared spectrum as well as the Raman spectrum. Now a much larger number of bands and intensity patterns are included in the assessment. The calculations should fit both experimental spectra well.
5. It is often the case that good fits in the 1000–3000 cm^{-1} region get poorer at low frequency and the very complex nature and high number of bands can make low-frequency fitting difficult. It may be necessary to scale the calculated energies in this region to get the best fit. If this is the case, recognition of specific bands whose energies are well known such as C—S and S—S stretches can add confidence. The authors treat this region with particular caution.

6. Effects such as resonance enhancement or changing pH make assignments from the calculation more difficult and care should be taken to ensure any assignments are correct. Resonance is dealt with in Chapter 4 but essentially some types of vibrations are selectively enhanced and some papers in the literature have made band assignments on resonantly enhanced spectra on the basis of observed intensity without considering which vibrations will give enhancement thus leading to improbable assignments.

Calculations reveal the true nature of vibrations and highlight the limitations of assignments made using simpler methods. For example, in Chapter 1 a vibration for benzene obtained from a DFT calculation is shown. It shows clearly what the vibration looks like and is exactly as expected. However, Figure 2.30 shows one vibration assigned previously as the azo stretch between the two nitrogens in the double bond which form the azo group. The calculation shows that there is significant movement on these nitrogens but there is significant movement on many other atoms too. This raises the question of how best to describe a vibration. For most purposes, the simple system set out in Chapter 1 and Section 2.10.1.3 is best. It is well established and allows easy communication. However, it should be borne in mind that this description of a vibration can be very approximate. If the bands are correctly assigned,

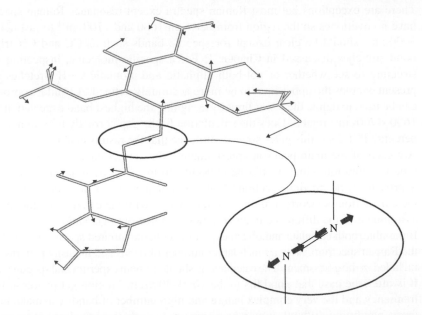

Figure 2.30. Diagram of the 'azo' stretch in an azo dye clearly showing that many more atoms are involved. This band was calculated at $1395\,cm^{-1}$ and found experimentally at $1373\,cm^{-1}$. The bond azo bond is circled and the direction of movement indicated for both directions in the expanded ellipse.

calculations give much more detail, which can be very informative but the difficulty is effective communication of so much information. How would you describe the vibration depicted in Figure 2.30 simply? In practice, the results are usually given as a table with the largest displacements from specific parts of the system given first. If a specific vibration is to be described in more detail it is usually shown with arrows showing the direction and relative amount of movement on each atom as shown in Figure 2.30. The arrows are shown in one direction but remember that these arrows mean movement in both directions away from and back through the rest position. To be clearly visible the length of the arrows is exaggerated. This can lead to mistaken conclusions. C—H vibrations often give very large displacements but in reality, the small size of the hydrogen atom puts it very close to the carbon atom reducing this greatly.

There have been many attempts at the development of artificial intelligence or expert systems to emulate the human thought process for spectral interpretation. Most have been successful for a limited range of similar chemical compounds or structures, but none have yet approached the full range attempted by humans. Many manufacturers of various spectroscopic instruments are including software training packages in their range of offerings, which very graphically demonstrate the fundamental principles of interpretation. Several now include excellent 3D graphic representation of band origins with simultaneous twisting, bending and stretching of bonds. These usually work very well for a limited range or group of molecules, but cannot be added to by the spectroscopist.

2.10.3 Spectra Formats for Transfer and Exchange of Data

Vibrational spectrocopists very quickly realized the potential of computers to manipulate spectra for quantitative or qualitative work. However, this initially required using software supplied by the instrument manufacturer or difficult and tedious manipulation of the spectra files for transfer to another computer. With the advent of PC workstations and the establishment of large commercial databases, the demand grew for a universal format for data transfer. In 1987, the Joint Committee on Atomic and Molecular Physical Properties (JCAMP) proposed a format to be used internationally. This is known as JCAMP-DX. The format was intended to represent all data in a series of labelled ASCII fields of variable length. Very quickly the major instrument manufacturers provided software to convert their spectra to/from JCAMP format. Unfortunately, whilst the data format was clearly specified, the file header format was less tightly specified. As a result, commas, spaces, etc. were used in different ways as delimiters. The effect was that each manufacturer supplied a slightly different JCAMP file. A number of commercial spectrum file converters are now available which allow for the import and export of files from most spectrometers into data-handling packages. The everyday use of 'cut and paste' techniques has also removed the need for file transfer as the image of a spectrum can easily be inserted into reports and presentations.

2.11 SUMMARY

The advantages and disadvantages of Raman spectroscopy from a practical viewpoint are very clear from what is said in this chapter. It is extremely flexible and can be configured in many different ways. The continued improvements in modern optics including small diode lasers, improved simple detectors and fibre-optic coupling have all led to the ability to use Raman scattering for problems for which we would not previously have considered it. Since it is a noncontact technique, it is possible to use it in a chemical factory with dust or inside the head of a combustion engine. Although the technique is limited by the fact that it is a weak effect, to some extent this can be overcome where the power density is high by the use of a microscope or particular forms of fibre optics. Thus, the future of Raman spectroscopy would appear to be set to advance particularly for specific analysis purposes. The disadvantage this creates for general use is that the range of choice requires an understanding of the subject and all the advantages cannot be obtained by purchasing one laser and a simple instrument. However, most laboratories find that modern Raman instrumentation – visible or NIR FT systems – can solve many of the standard problems for which Raman scattering is deemed to be a suitable technique.

REFERENCES

1. McCreery, R.L. (2000). *Raman Spectroscopy for Chemical Analysis*, Ch. 10. New York: Wiley.
2. Hendra, P., Jones, C., and Warnes, G. (1991). *FT Raman Spectroscopy*. Chichester: Ellis Horwood Ltd.
3. Bowie, B.T., Chase, D.B., and Griffiths, P. (2000). *Appl. Spectrosc.* **54**: 200A–207A.
4. Everall, N. and Lumsdon, J. (1991). *Vib. Spectrosc.* **2**: 257–261.
5. Pellow-Jarman, M.V., Hendra, P.J., and Lehnert, R.J. (1996). *Vib. Spectrosc.* **12**: 257–261.
6. Wang, H., Mann, C.K., and Vickers, J.V. (2002). *Appl. Spectrosc.* **56**: 1538–1544.
7. Chio, C.H., Sharma, S.K., Lucey, P.G., and Muenow, D.W. (2003). *Appl. Spectrosc.* **57**: 774–783.
8. Schrader, B. and Bergmann, G.Z. (1967). *Anal. Chem.* **225**: 230–247.
9. West, Y.D. (1996). *IJVS*, vol. 1, 1e, section 1. www.irdg.org/ijvs.
10. Hendra, P.J. (1996). *IJVS*, vol. 1, 1e, section 1. www.irdg.org/ijvs.
11. Chalmers, M. and Dent, G. (1997). *Industrial Analysis with Vibrational Spectroscopy*. London: Royal Society of Chemistry.
12. Dent, G. (1995). *Spectrochim. Acta A* **51**: 1975.
13. Chalmers, J. and Griffiths, P. (eds.) (2001). *Handbook of Vibrational Spectroscopy*, vol. 4, 2593–2600. New York: Wiley.
14. Dent, G. and Farrell, F. (1997). *Spectrochim. Acta A* **53**: 21–23.
15. Asselin, K.J. and Chase, B. (1994). *Appl. Spectrosc.* **48**: 699.
16. Petty, C. (1991). *Vib. Spectrosc.* **2**: 263.
17. Everall, N. (1994). *J. Raman Spectrosc.* **25**: 813–819.
18. Church, J.S., Davie, A.S., James, D.W. et al. (1994). *Appl. Spectrosc.* **48** (7): 813–817.

19. Louden, D. (1987). *Laboratory Methods in Vibrational Spectroscopy* (ed. H.A. Willis, J.H. van der Mass and R.J. Miller). New York: Wiley.
20. Rabolt, J.F., Santo, R., and Swalen, J.D. (1980). *Appl. Spectrosc.* **34**: 517.
21. Churchwell, J. and Bain, C. (2013). *Abstracts of Papers, 245th ACS National Meeting and Exposition*, New Orleans, LA (7–11 April 2013), PHYS-12.
22. Rentzepis, P., Dodson, R., and Taylor, C. (2017). *Abstracts of Papers, 254th ACS National Meeting and Exposition*, Washington, DC (20–24 August 2017), CHED-120.
23. Rentzepis, P., Dodson, R., and Taylor, C. (2017). *Abstracts of Papers, 254th ACS National Meeting and Exposition*, Washington, DC (20–24 August 2017), CHED-34.
24. Ngo, D. and Baldelli, S. (2016). *J. Phys. Chem. B* **120** (48): 12346–12357.
25. Littleford, R., Paterson, M.A.J., Low, P.J. et al. (2004). *Phys. Chem. Chem. Phys.* **6**: 3257–3263.
26. Lewis, J.R. and Griffiths, P.R. (1996). *Appl. Spectrosc.* **50**: 12A.
27. Angel, S.M., Cooney, T.F., and Trey Skinner, H. (2000). *Modern Techniques in Raman Spectroscopy* (ed. J.J. Laserna) Ch. 10. New York: Wiley.
28. Slater, J.B., Tedesco, J.M., Fairchild, R.C., and Lewis, I.R. (2001). *Handbook of Raman Spectroscopy*, Ch. 3 (ed. I.R. Lewis and H.G.M. Edwards), 41–144. New York: Marcel Dekker.
29. Song, L., Liu, S., Zhelyaskov, V., and El-Sayed, M.A. (1998). *Appl. Spectrosc.* **52**: 1364.
30. Schwab, S.D. and McCreery, R.L. (1987). *Appl. Spectrosc.* **41**: 126.
31. Xu, W., Xu, S., Lu, Z. et al. (2004). *Appl. Spectrosc.* **58**: 414–419.
32. Everall, N.J. (2010). *Analyst* **135**: 2512.
33. Everall, N.J. (2000). *Appl. Spectrosc.* **54**: 1515–1520.
34. Everall, N.J. (2000). *Appl. Spectrosc.* **54**: 773–782.
35. Macanally, G.D., Everall, N.J., Chalmers, J.M., and Smith, W.E. (2003). *Appl. Spectrosc.* **57**: 44.
36. Wood, B.R., Langford, S.J., Cooke, B.M. et al. (2003). *FEBS Lett.* **554**: 247–252.
37. Everall, N.J. (2014). *J. Raman Spectrosc.* **45**: 133–138.
38. McCabe, A., Smith, W.E., Thomson, G. et al. (2002). *Appl. Spectrosc.* **56**: 820.
39. Carter, D.A., Thompson, W.R., Taylor, C.E., and Pemberton, J.E. (1995). *Appl. Spectrosc.* **49**: 11.
40. Fountain, A.W. III, Mann, C.K., and Vickers, T.J. (1995). *Appl. Spectrosc.* **49**: 1048–1053.
41. NIST (2000). www.cstl.nist.goc/div837/Division/techac/2000/RamanStandards.htm (accessed 4 October 2018).
42. Kayser. www.kosi.com/raman/product/accessories/hca.html (accessed 4 October 2018).
43. Bowie, B.T., Chase, D.B., and Griffiths, P. (2000). *Appl. Spectrosc.* **54**: 164A–173A.

Chapter 3

The Theory of Raman Spectroscopy

3.1 INTRODUCTION

As shown in Chapter 1, the sharp pattern of bands which make up a spectrum makes it possible to use Raman spectroscopy for many types of analysis without a deep understanding of the nature of the effect. For example, it is possible to identify a molecule *in situ* from the pattern of bands and it may be possible to determine the amount of the compound which is present. However, a better understanding of the theory has real advantages. Much more information about a molecule and its surroundings can be obtained, the interpretation will be more secure, more possible pitfalls will be recognized and avoided and the background required to understand some of the more exciting modern developments will be understood. There are many more detailed books on the theory of Raman spectroscopy and some treatments can be quite extensive. This chapter selects out the topics required to explain the salient points needed for a thorough interpretation of a spectrum taken from a sample rather than an understanding of the scattering process per se. It is written to be as comprehensible as possible to spectroscopists with a wide range of backgrounds. For example, where a mathematical treatment is required to make a specific point, the key equations are explained without a full derivation and the physical information arising is described. The reader is referred to Refs. [1–7] for a more thorough coverage.

The usual approach to Raman scattering theory starts with the theory of light scattering based on a paper by Mie in 1908 [8]. It can be developed to demonstrate that Stokes and anti-Stokes scattering should arise and it will give the fourth power law of scattering. However, Raman scattering is a very good example of the quantum effect and hence this approach, often used in textbooks and of importance in understanding scattering per se, does not directly help interpret a spectrum or help

Modern Raman Spectroscopy: A Practical Approach, Second Edition. Ewen Smith and Geoffrey Dent.
© 2019 John Wiley & Sons Ltd. Published 2019 by John Wiley & Sons Ltd.

understand how modern computer programs calculate peak intensities. It is not described further here. Accounts of the theory can be obtained in Refs. [1–3]. The main focus of the chapter is on understanding polarizability (α), the molecular property which defines Raman intensities and the insights this gives to understanding to Raman scattering.

3.2   ABSORPTION AND SCATTERING

When light interacts with matter, it can be absorbed or scattered. The process of absorption, discussed briefly in Chapter 1, requires that the energy of the incident photon corresponds to the energy gap between the ground state of a molecule and an excited state. It is the basic process used in a wide range of spectroscopic techniques and will be familiar to many readers. In contrast, scattering can occur whether or not there is a suitable pair of energy levels to absorb the radiation, and the description of the interaction between the radiation and the molecule requires a different approach.

When a light wave, considered as a propagating oscillating dipole, passes over a molecule, it can interact and distort the cloud of electrons round the nuclei. In the visible region, the wavelength of the light is between about 400 and 700 nm and a small molecule such as carbon tetrachloride is about 0.3–0.4 nm. The amplitude of the oscillating dipole from the light which is at 90° to the propagating direction is much larger than the size of a molecule. If the light interacts with the molecule, it causes the electrons to polarize and go to a short-lived higher energy arrangement from which it is reradiated. At the instant the higher energy arrangement forms, the energy present in the light wave is transferred into the molecule. This results in a molecule with a different electron geometry but with a lifetime that is so short the nuclei do not have time to move to the lowest energy position for that geometry. This arrangement is not a true state of the molecule and is called a virtual state. The actual shape of the distorted electron arrangement will depend on the molecule and on how much energy is transferred to it and hence is dependent on the frequency of the laser used. In addition, the frequency of the laser defines the energy of the virtual state.

The scattering process differs from the absorption process in a number of ways. Firstly, the additional energy does not promote an electron to any one excited state of the static molecule; all states of the static molecule are involved to different extents and are mixed together to form the virtual state. The radiation is scattered in different directions and not absorbed by the molecule. Lastly, and this will be dealt with later in this chapter, there is a link between the polarization direction of the exciting and scattered photons, which can be of value in assigning particular vibrations.

Two types of scattering are readily identified. The most intense form of scattering, Rayleigh scattering, occurs when the virtual state relaxes without any nuclear movement. This is essentially an elastic process with no appreciable change in energy. Raman scattering is a rarer event, which involves only one in 10^6–10^8 of the photons scattered. This occurs when on excitation, the electrons interact with the nuclei and some energy is transferred. If the molecule is in the ground state initially, the photon scattered from the virtual state is lower in energy by the amount transferred to the nuclei and the molecule starts to vibrate. This is Stokes scattering. However, if the molecule is in a vibrationally excited state, the scattering process returns the molecule to the ground state and the scattered radiation is higher in energy than the excitation frequency by the amount of the vibrational energy added. This is anti-Stokes scattering. Although the virtual state is very short lived, in Stokes scattering, energy has been transferred so that nuclear movement will carry on after the electrons return to their equilibrium position. Figure 1.2 in Chapter 1 shows a simple diagram illustrating Rayleigh and Raman scattering. In each case the energy of the virtual state is defined by the energy of the incoming laser.

Molecules at room temperature and before excitation are most likely to be in the ground vibrational state. Therefore, the majority of Raman scattering will be Stokes Raman scattering. The ratio of the intensities of the Stokes and anti-Stokes scattering is dependent on the number of molecules in the ground and excited vibrational levels. This can be calculated from the Boltzmann equation,

$$\frac{N_n}{N_m} = \frac{g_n}{g_m} \exp\left[\frac{-(E_n - E_m)}{kT}\right] \tag{3.1}$$

N_n is the number of molecules in the excited vibrational energy level (n),
N_m is the number of molecules in the ground vibrational energy level (m),
g is the degeneracy of the levels n and m,
$E_n - E_m$ is the difference in energy between the vibrational energy levels,
k is Boltzmann's constant (1.3807×10^{-23} J K^{-1}).

We shall see, when we consider symmetry later in this chapter, that some vibrations can occur in more than one way and the energies of the different ways are the same, so that the individual components cannot be separately identified. The number of these components is called the degeneracy and is given by the symbol g in Eq. (3.1). Since the Boltzmann distribution has to take into account all possible vibrational states, we have to correct for this. For most states g will equal 1 but for degenerate vibrations it can equal 2 or 3.

3.3 STATES OF A SYSTEM AND HOOKE'S LAW

The next three sections provide information to inform the interpretation of a spectrum. In this section, the basic way in which the energy of specific vibrations can be estimated is described and in the Sections 3.4 and 3.5 the way in which intensities can be estimated are described.

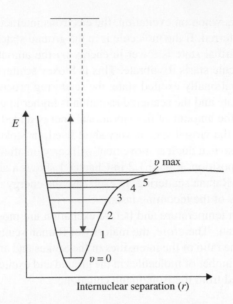

E

v max

5
4
3
2
1
v = 0

Internuclear separation (*r*)

Figure 3.1. A typical Morse curve for an electronic state showing the fundamental and over-tone levels for one vibration as horizontal tie lines. Raman excitation and scattering processes are shown in red. The dotted lines are to show that the length of these lines in practice may be longer.

Any molecule consists of a series of electronic states, each of which contains a large number of vibrational and rotational states. In Figure 3.1 a sketch of a typical ground electronic state of a molecule is shown for one vibration and with no rotational levels. The *y*-axis represents the energy of the system and the *x*-axis the internuclear separation. The curved line represents the electronic state. At large inter-nuclear separations, the atoms are essentially free and as the distance decreases they are attracted to each other to form a bond. If they approach too closely, the nuclear forces cause repulsion and the energy of the molecule rises steeply as shown. Thus, the lowest energy is at the bond length. However, within the curve, not every energy is possible since the molecules will be vibrating and the vibrational energies are quantized. The tie lines are the quantized vibrational states. A vibrational state of a specific electronic state is often called a vibronic state and this term is used in fol-lowing sections.

What is shown in the figure refers to one vibration. The first level (*v* = 0) is the ground state and the second level (*v* = 1) is the first excited state. One quan-tum of the energy difference between the two can be absorbed causing the mol-ecule to vibrate. The levels above this require energies of approximately but not exactly two times, three times, four times, etc. of the quanta required to move the molecule from the ground state 0 to the first excited state 1. The reason the differences are not exact is that the curve shown is not exactly a parametric oscillator. When absorption occurs involving a change of more than one quantum, the band obtained is called an overtone. As we shall see, in Raman scattering this occurs only in special circumstances such as with resonance Raman scattering described in the next chapter. In most Raman spectra overtones are very weak or nonexistent. To describe all the possible vibrational states in a larger molecule in the way shown in Figure 3.1, a similar set of tie lines but at different energies is required for each vibration. Further, vibrations can combine so that one quantum of one vibration and one of another vibration will give a new level. In the spectrum, peaks due to these combinations are called combination bands and like overtones appear only in certain circumstances. To make matters even more complicated, rotational levels, which are of lower energy than vibrational levels, also require to be added and rotational progressions are built on each vibronic level. A diagram with all these levels is too complex to use and conventionally is simplified either by showing all the levels for one vibration as in Figure 3.1 or the ground state and the first excited state for each vibration of the molecule. In Raman scattering the exciting radiation defines the virtual state. Because any one Raman scattering event occurs on a faster time scale than the time to complete one vibrational cycle, there is only a very small change in nuclear geometry and arrows showing a scattering event should be vertical.

There are features in Figures 1.2 and 3.1 which are potentially misleading. For example, if excitation at 500 nm is used this is equivalent to 20000 cm^{-1} whereas single and double carbon to carbon bond vibrations are below 1600 cm^{-1}. Particularly for low frequency vibrations, the red vertical lines in Figure 3.1 should be longer. However, the energy of the laser radiation is not represented to scale because the desire is to show the vibrational spacing clearly and plotting the true excitation energy would lead to a very large separation between the ground state and the virtual state with very small spaces between the vibronic levels. Again, for clarity the difference in the energy separation between the vibronic levels is exaggerated and is normally much smaller in real systems. In addition, since the vibrational movement during the scattering event is small, arrows representing the excitation and emission processes should lie virtually on top of each other. They are separated in Figure 3.1 for clarity.

The shape of the Morse curve makes it somewhat complex to calculate the energy of vibronic levels and so simple theory uses the harmonic approximation. In this approach, for a diatomic molecule, the Morse curve shown is replaced by a parabola calculated by considering the molecule as two masses that are connected by a vibrating spring. With this approach, Hooke's law (Eq. 3.2) gives the relationship between frequency, the mass of the atoms involved in the vibration and the bond strength.

$$\nu = \frac{1}{2\pi c}\sqrt{\frac{K}{\mu}} \qquad\qquad (3.2)$$

where c is the velocity of light, K is the force constant of the bond between A and B and is a measure of bond strength. μ is the reduced mass of atoms A and B of masses M_A and M_B:

$$\mu = \frac{M_A M_B}{M_A + M_B} \tag{3.3}$$

Hooke's law makes it easy to understand the approximate order of the energies of specific vibrations. The lighter the atoms, the higher the frequency will be. Thus C—H vibrations lie just below and just above $3000\,cm^{-1}$ for aliphatic and aromatic systems, respectively, and C—I vibrations are at less than $500\,cm^{-1}$. The force constant is a measure of bond strength. The stronger the bond, the higher the frequency will be so a —C=C— stretch will be higher in energy than a —C—C— stretch. A table of vibrational energies for different moieties is given in Chapter 1.

Two other points should be noted. The harmonic approximation predicts that the overtones of a molecule will be equally spaced but the departure from harmonicity in a real system means that the energy separations between levels will decrease as shown in Figure 3.1. Secondly, the distribution of the electron density along each tie line is of importance in working out the efficiency of the resonance Raman process. Basically, the maximum electron density in the ground state is in the middle of the tie line and it moves further to both sides as the vibrational quantum number ν increases. We will make use of this later in discussing resonance in Chapter 4.

3.4   THE BASIC SELECTION RULE

The basic selection rule is that Raman scattering arises from a change in polarizability in the molecule and only one quantum unit change is possible (i.e. $\Delta\omega = \pm1$). As we shall demonstrate later, symmetric vibrations will give the most intense Raman scattering. This is in contrast to infrared absorption where a dipole change gives intensity and this means asymmetric rather than symmetric vibrations will be intense. The same $\Delta\omega = \pm1$ selection rule applies to both Raman scattering and infrared absorption but the Raman selection rule tends to be more rigorous and except in certain circumstances such as where resonance is present, no overtones are usually seen. This can be an important piece of information. CC bonds except triple bonds occur below about $1650\,cm^{-1}$ and C—H stretches occur above about $2800\,cm^{-1}$ and usually there are many fewer bands between 1650 and $2800\,cm^{-1}$. Often they are easy to be assigned to specific groups such as a triple CC bond or a cyanide. One aspect of this of growing importance is where assignments are made from computer-predicted spectra (see Chapter 2). A simple check to see if the prediction is reasonable is to check whether the gap is predicted correctly. In addition, specific groups with the structure to give a vibration in the gap region are used as tags in some applications because they are so easily identified.

3.5 NUMBER AND SYMMETRY OF VIBRATIONS

With any molecule, the energy can be divided into translational energy, vibrational energy and rotational energy. Translational energy can be described in terms of three vectors at 90° to each other and so has three degrees of freedom. Rotational energy for most molecules can also be described in terms of three degrees of freedom. However, for a linear molecule there are only two rotations. The molecule can either rotate around the axis or about it. Thus, molecules have three translational degrees of freedom and three rotational degrees of freedom except for linear molecules which have three translational degrees of freedom and two rotational degrees of freedom. All other degrees of freedom will be vibrational degrees of freedom and each is equivalent to one vibration. Therefore, the number of vibrations to be expected from a molecule with N atoms is $3N - 6$ for all molecules except linear systems where it is $3N - 5$.

From this it is possible to work out the number of vibrations which occur. However, this does not mean the vibrations are either Raman or infrared active and from selection rules we would not expect to observe all vibrations in either spectroscopy.

As discussed in Chapter 1, for a simple diatomic molecule, which by definition is linear, there is one vibration. For a simple homonuclear diatomic like oxygen or nitrogen this is a symmetric vibration in which we would not expect any infrared activity, but, since the bond is stretched, we would expect a change in polarizability to occur. Thus, one band would be expected in the Raman spectrum and there would be no band in the infrared spectrum. However, in heteronuclear diatomic there would be a dipole and a polarization change, so we would expect a one band in both Raman and infrared spectra.

In a more complex molecule, when a molecule has a number of symmetry elements in its structure, more selection rules apply. Consider a square planar molecule such as $AuCl_4^-$ for which selected vibrational movements are indicated by arrows in Figure 3.2. This molecule is said to have a centre of symmetry. The definition of a centre of symmetry is that any point in the molecule reflected through the central point will arrive at an identical point on the other side. Thus, in this molecule, ignoring vibrational

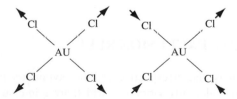

Figure 3.2.   Illustration of two vibrations in the centrosymmetric ion $AuCl_4^-$. The arrows represent direction of motion. It should be noted that during a vibration the atoms will all reverse direction later to complete the vibrational cycle.

movement, any chlorine atom reflected through the gold centre will arrive at an identical chlorine atom on the other side. An example of a molecule that does not possess this property is the nitrate ion in which an oxygen reflected through the nitrogen in the centre would arrive at a point in space.

We need to have some way of describing the vibrations in a molecule. In principle, it would be possible to use x, y and z coordinates for each atom and simply explain how each atom moves by how much these coordinates change. This would be complicated and we would not understand the nature of the information readily. The usual way is to use normal coordinates as shown for $AuCl_4^-$ in Figure 3.2. Normal coordinates of a molecule make use of the natural directions of bonds and are those coordinates in which all atoms vibrating go through the centre of gravity of the molecule at the same time. The value of normal coordinates is that they provide a much better visual pattern of what a vibration looks like. Two vibrations for the $AuCl_4^-$ molecule using normal coordinates are illustrated in Figure 3.2.

In one vibration, all the atoms move out at the same time, and in the other, three atoms move in as one moves out. The arrows representing the movement of atoms can also be reflected through the centre in the same way as atoms. The assignment of the molecule as centrosymmetric was based on the properties of the atoms with the molecule at rest in its equilibrium position. The vibrational movements do not affect that but also have symmetry properties. Clearly there is a difference between the two vibrations as shown in Figure 3.2. Vibrations of the first type are symmetric and are called even, or gerade, and are labelled g, whereas those of the second type are asymmetric and called odd, or ungerade, and are labelled u. This labelling applies only to molecules with a centre of symmetry. It will not, for example, apply in nitrate.

For any molecule with symmetry elements, it is possible to use symmetry to help predict which vibrations will be strong in Raman scattering by applying group theory. The approach is very powerful in experiments involving molecules of high symmetry, and, in addition, the use of labels which arise from group theory is common throughout the literature. The basics of this approach are described later in this chapter. However, it is often not useful in experiments with more complex molecules and so an extensive treatment is outside of the scope of this book. Good texts such as the book by Cotton [3] describe the application of symmetry in detail.

3.6 THE MUTUAL EXCLUSION RULE

One important result is that, irrespective of other symmetry considerations, for a centrosymmetric molecule, only vibrations which are g in character can be Raman active and only vibrations which are u in character can be infrared active. Therefore, in a centrosymmetric molecule no band can be both Raman and infrared active. This is called the mutual exclusion rule. As we will see later, Raman scattering can be described in terms of expressions in which the key elements are integrals coupling the ground vibronic state and the excited state through an operator. The *g* and *u* labels on each state and the operator can be multiplied out and the final product must contain the totally symmetric representation which is *g* in character. The rules are $g \times g = g$, $u \times u = g$ and $g \times u = u$. The next section discusses an expression to derive polarizability, the numerator of which consists of two triple integrals multiplied together (Eq. 3.5). Given that the ground state is *g* and the operators are *u*, the final state must be *g* to be allowed. In contrast, the infrared operator is *u* in character and only one integral is involved so the excited state must be *u* if the vibration is to be allowed. In molecules without a centre of symmetry, there is no such specific rule and both symmetric and asymmetric vibrations can appear. Nonetheless, in general, symmetric vibrations are more intense in Raman scattering and asymmetric vibrations in infrared scattering.

3.7 UNDERSTANDING POLARIZABILITY

So far, fairly standard arguments have been used to show how to predict the energy and intensity of the bands found in the spectrum. However, there are good reasons for going deeper into the theory. The reason for the $\Delta \omega = \pm 1$ selection rule becomes apparent, a deeper understanding of the Raman scattering process is obtained, the changes found with resonance Raman scattering can be explained, more accurate intensities can be calculated and some insight into the methods used to obtain intensities in computer predicted spectra is obtained. The salient points can be obtained by analysing two key equations.

The first of these, Eq. (3.4) is obtained from light scattering theory and gives the intensity of Raman scattering:

$$I = K l \alpha^2 \omega^4 \tag{3.4}$$

K consists of constants such as the speed of light, *l* is the laser power, ω the frequency of the incident radiation and α the polarizability of the electrons in the molecule. Thus, two of the parameters which are variable are under the control of the spectroscopist, who can set the laser power (*l*) and the frequency of the incident light (ω). The way in which these can be used to maximize the intensity of Raman scattering has already been considered (Chapter 2).

A second equation is required to understand the role of the molecular property, the polarizability α. The equation used here to describe polarizability in the molecule is known as the Kramer Heisenberg Dirac (KHD) expression. There are other approaches such as time-dependent theory, but this approach is commonly used and it or the conclusions from it appear in many papers on Raman scattering. It is a large equation but it can be easily understood with little in the way of mathematical

knowledge. All the symbols are defined below and the process being described is the
one shown diagrammatically in Figures 1.2 and 3.1.

$$(\alpha_{\rho\sigma})_{GF} = k\sum_{I}\left(\frac{\langle F|r_{\rho}|I\rangle\langle I|r_{\sigma}|G\rangle}{\omega_{GI} - \omega_{L} - i\Gamma_{I}} + \frac{\langle I|r_{\rho}|G\rangle\langle F|r_{\sigma}|I\rangle}{\omega_{IF} + \omega_{L} - i\Gamma_{I}}\right) \tag{3.5}$$

α is the molecular polarizability and ρ and σ are the incident and scattered polarization directions. Σ is the sum over all vibronic states of the molecule as might be
expected from the nonspecific nature of scattering. k is a constant. G is the ground
vibronic state, I a vibronic state of an excited electronic state and F the final vibronic
state of the ground state. G and F are simply the initial and final states of the Raman
scattering process. We will consider the numerator and the denominator separately
and define the terms in the denominator in due course.

The numerator begins the description of the virtual state by mixing all excited
electronic state vibronic states with the ground and final states. To understand this,
consider only the first term. It consists of two integrals but this may not be immediately obvious to some. Because of the complexity of Eq. (3.5) it is usual to write the
integrals using 'bra' and 'ket' ($\langle|$ and $|\rangle$) nomenclature rather than standard integrals.
The integral $\langle I|r_{\sigma}|G\rangle$ is similar to those used in electronic adsorption spectra to
describe the absorption process but in that case, light is promoted to a specific excited
state of the molecule. In Raman scattering the requirement is to describe scattering
using all vibronic states of the molecule and each integral is best considered as a mixing of the ground state and an excited state. The sum (Σ) of all the upward terms will
describe the distorted electron configuration in the complex between the molecule
and the light (i.e. the virtual state). Starting from the right-hand edge of the expres-
sion, $|G\rangle$ is a wave function representing the ground vibronic state of the ground
electronic state. The operator r_{σ} is the dipole operator and the mathematical process
of it operating on $|G\rangle$ and multiplying the product with the excited state $\langle I|$ mixes
the two states. As a simple example, consider a molecule which is spherical in the
ground state. This can be represented as an s orbital. When radiation interacts with
it to form a virtual state the electrons form an ellipse. This can be represented very
approximately by mixing the s orbital ground state with a p orbital excited state or
<p|r_{0}|s>. This would be improved by adding other excited states (Figure 3.3).

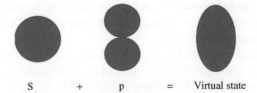

S   +   p   =   Virtual state

Figure 3.3.   A very simple example to explain the process indicated by the integral. A ground
state s is mixed with an excited state p to give an elliptical virtual state.

The left-hand integral describes the scattering process from the virtual state to leave the molecule in the final state $\langle F|$. Thus, the first of the two triple integrals mixes a ground and an excited state and the second of these integrals mixes the excited state and the final state. In Raman scattering one of these double integrals is needed for each vibronic state I and the whole is summed as shown by the Σ in front of this expression. Since it is a mixing between two states, which is being described, there is no reason why this process should start in the ground state. Thus, in the second term in Eq. (3.5), an equivalent expression to that in the first term is added. This starts with the excited states and mixes the excited and ground states together. Fortunately, as we shall see in the next paragraph, this term is less significant in Raman scattering.

Some excited states are much closer in energy to the virtual state than others and those nearest in energy might be expected to contribute the most to its structure. This is included in Eq. (3.5) in the denominator. As long as the energy of the laser and an excited state are not very similar, the term $i\Gamma_I$ is small compared to ω_{GI}, the energy difference between the ground state and an excited vibronic state. In the first term, since ω_L is subtracted from ω_{GI}, the nearer a specific excited state I is in energy to ω_L, the smaller the denominator will be and the larger part the particular expression in the numerator for that state will play in the final expression. Further, because ω_{GI} and ω_L are added and not subtracted in the second term of Eq. (3.4), the denominator will always be large. Consequently, the second term plays a smaller role in describing the polarization process and will now be neglected. Without $i\Gamma_I$, when ω_L is the same energy as that of a specific electronic transition, then the denominator of the first term would go to zero and the result would be that the scattering would become infinite! The term $i\Gamma_I$ relates to the lifetime of the excited state and affects the natural breadth of Raman lines so, although it is small, it is a vitally important part of the basic equation.

Thus, the magnitude of each of the expressions in the numerator of term 1 will depend on the nature of the states but the contribution to the final sum will be weighted by the denominator so that those closest in energy to the exciting radiation will contribute proportionately more.

In this paragraph, we further analyse the KHD expression particularly to understand the selection rules in Raman scattering and to lay the foundation to understand resonance Raman scattering explained in Chapter 4. To do this, the states are usually split up into electronic and vibrational components using the Born Oppenheimer approach. In this approach, the total wave function is split up into separate electronic (θ), vibrational (Φ) and rotational (r) components.

$$\Psi = \theta \cdot \Phi \cdot r \tag{3.6}$$

This is a very successful way of approaching many spectroscopy problems. It works because there is a significant difference in the timescale of electronic, vibrational and rotational transitions, so treating each separately is a reasonable approximation. The very light electrons involved in a pure electronic transition will change from a ground to an excited state in a timescale in which there is very little movement of the nucleus (10^{-13} or less of a second). Vibrational transitions occur in about 10^{-9} of a second and are faster than rotational transitions. Although rotational states can be of interest, particularly in gas-phase Raman spectra, for the purposes of the limited theory given here, the rotational contribution will be neglected. To readers interested in the theory in more depth, this is an interesting approximation to use to describe Raman scattering where the interaction of electronic and vibrational states is at the heart of the process but it works as explained below.

Using this approach, the electronic and vibrational terms can be treated separately. θ is the electronic part of the expression and will depend on both the nuclear and electronic coordinates (R and r, respectively) whereas the vibrational part Φ, which involves displacement of the heavier nuclei, will depend entirely on the nucleic coordinates (R). This separation allows the vibrational and electronic functions in the numerator in the KHD expression to be split up.

$$\langle I|r_\sigma|G\rangle = \langle \theta_I \cdot \Phi_I|r_\sigma|\theta_G \cdot \Phi_G\rangle = \langle \theta_I|r_\sigma|\Phi_G\rangle\langle \Phi_I | \Phi_G\rangle \tag{3.7}$$

As stated earlier, the Raman process is so fast that during the lifetime of the virtual state the nuclear movement only has time to begin and only a small change in nuclear geometry occurs. This means that the electronic part of the wave function can be approximated to what happens when the nuclei are at rest with a correction term to allow for the change in electronic structure as the nuclei move. To make this a little simpler, the electronic term from the expression above is written as

$$\langle \theta_I | r_\sigma | \theta_G\rangle = M_{IG}(R) \tag{3.8}$$

The movement is described by a Taylor series with the value at rest being the first and largest term $M_{IG}(R_0)$ where R_0 represents the coordinates at the equilibrium position. The second and higher terms describe the effect of movement along a particular coordinate R_ε and even the second term is relatively small. Thus, all but the first and second terms can be neglected. For simplicity, the first and second terms are written as M and M':

$$M_{IG}(R) = M_{IG}(R_0) + \left[\frac{\delta M_{IG}}{\delta R_\varepsilon}\right]_{R_0} R_\varepsilon + \text{higher order terms} \tag{3.9}$$

In this way, the KHD expression can be solved. We will not attempt the mathematics here but there are more details in Ref. [6]. Carrying out this procedure leads to the equation below. It looks complex but can easily be simplified.

$$\left(\alpha_{\rho\sigma}\right)_{GF} = kM_{IG}^2 \left(R_0\right)\sum_I \frac{\left\langle\Phi_{R_F}\mid\Phi_{R_I}\right\rangle\left\langle\Phi_{R_I}\mid\Phi_{R_G}\right\rangle}{\omega_{GI}-\omega_L-i\Gamma_I} \quad \text{(A-term)}$$

$$+kM_{IG}(R_0)M'_{IG}(R_0)\sum_I \frac{\left\langle\Phi_{R_F}\mid R_\varepsilon\mid\Phi_{R_I}\right\rangle\left\langle\Phi_{R_I}\mid\Phi_{R_G}\right\rangle+\left\langle\Phi_{R_F}\mid\Phi_{R_I}\right\rangle\left\langle\Phi_{R_I}\mid R_\varepsilon\mid\Phi_{R_G}\right\rangle}{\omega_{GI}-\omega_L-i\Gamma_I} \quad \text{(B-term)}$$

$$(3.10)$$

The two terms shown in the equation are known as A-term and B-term. Outside the summation sign, there is a term corresponding to the electronic component of the Raman scattering (M). It is squared in A-term and is M times the much smaller correction factor M' in B-term. Thus, this part of the expression is much larger in A-term than in B-term.

In A-term, the numerator inside the summation sign consists simply of a multiplication of all possible vibrational wave functions. There is a theorem called the closure theorem which states that when all vibrational wave functions are multiplied together, the final answer is zero. Thus, no Raman scattering will be obtained from A-term. In B-term, an operator, the coordinate operator R_ε, is present in the numerator. This operator describes the effect of movement along the molecular axis during the vibration. One feature of this operator is that the integral will only have a finite value when there is one quantum of energy difference between the initial state on which it operates and the final state. This means that only vibrations between levels one quantum of energy apart will give Raman scattering – the basic selection rule stated earlier. Further, this means no overtones should occur in Raman scattering and this is a good selection rule. Overtones are not seen unless there is some form of special effect such as resonance.

3.8 POLARIZABILITY AND THE MEASUREMENT OF POLARIZATION

When radiation is emitted from a source, a number of photons are emitted and each photon consists of an oscillating dipole. Observed at 90° to the direction of propagation, the beam looks like a wave. Observed along the line between the observer and the light source, each photon will appear as a line caused by the oscillating dipole. In general, the angle of the line to the observer from a light source like a bulb is random, but by passing the light through a suitable optical element such as a Nicol prism or a piece of Polaroid film, all the dipoles can be made to propagate in one direction. This is called plane or linearly polarized radiation. The lasers that are normally used for excitation in Raman scattering are at least partially polarized. Many Raman spectrometers also have an optical element, a polarizer, that can be put in the beam to ensure that the light is linearly polarized and to alter the direction of the polarization if required. It is not necessary for the exciting beam to be linearly polarized to obtain good Raman spectra but if further analysis as discussed below is to be undertaken it is required.

When a photon interacts with the molecule, the electron cloud is distorted by an amount that depends on the ability of the electrons to polarize (i.e. the polarizability, α). The light causing the effect is polarized in one plane, but the effect on the electron cloud is in all directions. This can be described as a dipole change in the molecule in each of the three Cartesian coordinates x, y and z. Thus, to describe the effect of the beam on the electron cloud, three dipoles require to be considered. The simple expression is that a dipole μ is created in the molecule by the field from the incident photon E, the magnitude of which is also dependent on the polarizability α.

$$\mu = \alpha E \tag{3.11}$$

and there will be three of these μ_x, μ_y and μ_z. To describe the relationship between the polarization angle of the linearly polarized light and the molecule, the polarizability components of the molecule are usually labelled, an example of which is shown below:

$$\alpha_{xx}$$

The first subscript x refers to the polarizability of the molecule in the x direction, and the second x refers to the direction of polarization of the incident light.

The dipole created in the x direction is given by

$$\mu_x = \alpha_{xx}E_x + \alpha_{xy}E_y + \alpha_{xz}E_z \tag{3.12}$$

Similar expressions give μ_y and μ_z.

Thus, the polarizability of the molecule is a tensor,

$$\begin{bmatrix} \mu_x \\ \mu_y \\ \mu_z \end{bmatrix} = \begin{bmatrix} \alpha_{xx} & \alpha_{xy} & \alpha_{xz} \\ \alpha_{yx} & \alpha_{yy} & \alpha_{yz} \\ \alpha_{zx} & \alpha_{zy} & \alpha_{zz} \end{bmatrix} \begin{bmatrix} E_x \\ E_y \\ E_z \end{bmatrix} \tag{3.13}$$

There are specific advantages of this rather complex arrangement. In Raman scattering the incident and scattered beams are related. If radiation of a particular polarization is used to create the Raman scattering, the polarization of the scattered beam is related to but not necessarily the same as that of the incident beam. For this reason, a Raman spectrometer may have an optical element, the polarizer, to control the polarization of the incident beam. It ensures that the radiation is plane polarized and determines the angle of the plane of the incident radiation. A second element, the analyser, analyses the polarization of the scattered beam. The analyser works by allowing the polarized light to pass through to the detector only in one plane. It can be set to allow transmission of scattered radiation only in the plane of the incident radiation (called parallel scattering) as shown in Figure 3.4. It can then be rotated 90°

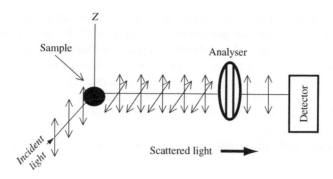

Figure 3.4. Parallel arrangement to monitor polarization. The double arrows indicate the direction of the oscillating dipole. The linearly polarized incident light from the laser is cleaned up by the polarizer as shown by the double arrows. After scattering, both polarizations are present and the analyser transmits only the polarization in the same direction as that of the incident light. The analyser can be rotated 90° to transmit only light rotated by the scattering process (perpendicular scattering).

to allow light in which the polarization direction has been changed by the scattering process to pass through to reach the detector (called perpendicular scattering).

If a single crystal is mounted on a goniometer or similar device the polarization direction of the incident radiation can be set to lie along each crystal axis in turn. This works best for crystals in higher symmetry space groups but not cubic. Light is a dipole property, which means that it can be resolved into three components at right angles, one along the crystal in the direction of the light and the other two at 90° to each other in a plane perpendicular to it. In some higher symmetry space groups such as the tetragonal space group, the optical and crystal axes can be aligned. Light polarized in the z direction, passing through the crystal with the dipole parallel to the z-axis and with a molecular axis parallel to it, will pick out the component α_{zz}. By rotating the sample, other individual terms can be obtained. Light, which is not sent down an axis of the crystal, will rotate within it and in many crystal space groups, the crystal axes are not at right angles and bear a complex relationship to the molecular axes. The technique can still be used but the analysis is quite complex. This approach is very informative but straightforward only for a limited number of crystals. It is not dealt with further here but the reader should be aware of the possibility that, when using single-crystal samples or other samples with oriented molecules, the intensities of the bands may be affected by the angle of the incident beam to the sample.

Often, samples examined are either in the gas phase or in solution. In either case there is no ordering of the axes of the molecule to the polarization direction of the light but information can still be obtained from polarization measurements. For samples such as this, it is useful to express the average polarizability in terms of two separate quantities that are invariant to rotation, namely isotropic and anisotropic scattering. Isotropic scattering is measured with the analyser parallel to the plane of the incident radiation and anisotropic scattering with the analyser perpendicular to the plane. The ratio of anisotropic scattering to isotropic scattering is called the depolarization ratio. It is possible to solve the tensor shown in Eq. (3.13) and calculate this ratio (see Ref. [1]). This is called the depolarization ratio (ρ). Here, we illustrate the salient equations but do not give details since this ratio is usually used qualitatively and is often talked about but seldom calculated. It is also dependent on the instrument geometry.

The isotropic and anisotropic parts of the tensor are represented in Eqs. (3.14) and (3.15),

$$\bar{\alpha} = \frac{1}{3}(\alpha_{xx} + \alpha_{yy} + \alpha_{zz}) \tag{3.14}$$

$$\gamma^2 = \frac{1}{2}\left[\left(\alpha_{xx} - \alpha_{yy}\right)^2 + \left(\alpha_{yy} - \alpha_{zz}\right)^2 + \left(\alpha_{zz} - \alpha_{xx}\right)^2 + 6\left(\alpha_{xy}^2 + \alpha_{xz}^2 + \alpha_{yz}^2\right)\right] \tag{3.15}$$

and the effect on parallel and perpendicular polarization of Eqs. (3.7) and (3.8),

$$\bar{\alpha}_{\parallel}^2 = \frac{1}{45}(45\bar{\alpha}^2 + 4\gamma^2) \tag{3.16}$$

$$\bar{\alpha}_{\perp}^2 = \frac{1}{15}\gamma^2 \tag{3.17}$$

This gives a ratio between parallel and perpendicular scattering as

$$\rho = \frac{\bar{\alpha}_{\wedge}^2}{\bar{\alpha}_{\parallel}^2} = \frac{3\gamma^2}{45\bar{\alpha}^2 + 4\gamma^2} \tag{3.18}$$

The importance of this information for a molecule with appreciable symmetry in solution or in the gas phase is that the depolarization ratio depends on the symmetry of the vibration. Symmetric vibrations have the lowest depolarization ratios. Thus, measurement of parallel and perpendicular scattering using the analyser to obtain the depolarization ratio for a molecule with symmetry provides a check on assignments of specific bands in the spectrum which is not dependent on the initial assignment made from intensity and predicted frequency. This check is not available with absorption spectroscopies such as infrared.

There is one final practical point which has to be borne in mind. When radiation from the analyser is detected via a monochromator, the efficiency of the grating used to split up the light is dependent on the plane of polarization. This means the grating will transmit radiation to the detector more efficiently for either parallel or perpendicular polarization and consequently the apparent depolarization ratio will be wrong. The most conventional way to overcome this problem is to add an extra element, a scrambler, which scrambles the polarization of the light before it enters the monochromator so that the detector is equally efficient for all polarization directions of the incoming radiation. There are other ways of doing this. For example, a half-wave plate can be inserted instead. This rotates the light by 90° and is swung into the beam in only one direction of the analyser so that the light in both analyser positions enters the monochromator in the same direction.

Failure to appreciate this effect can be serious. Laser radiation is usually linearly polarized to a significant extent. In the absence of polarizers and analysers and if no scrambler is in place, the laser acts as the polarizer and the monochromator as the analyser. Commercial systems are set up to give maximum throughput so that the polarization direction of the laser is aligned to give maximum scattering from the grating and for most larger molecules, this does not affect the spectrum appreciably, but if a laser polarized at 90° is added and the molecule has symmetry, the effect can be considerable.

3.9   SYMMETRY ELEMENTS AND POINT GROUPS

Many large organic molecules have very little symmetry and many papers simply ignore symmetry implications. However, when present, symmetry can have a significant effect on the spectrum and is essential to interpret the spectrum of small molecules such as carbon tetrachloride or inorganics like $AuCl_4^-$. There are many examples of larger molecules with symmetry which does not always need to be exact to produce an important effect. For example, azobenzene is centrosymmetric and the mutual exclusion rule will apply. Phthalocyanines and porphyrins have essentially but normally not exactly D_{4h} symmetry and many other molecules can have at least part of the molecule in a high symmetry configuration. The section below sets out some of the basics allowing the spectroscopist to understand the text in many articles and to recognize symmetry considerations in their own studies. For those more seriously involved, more specialist texts will be required [3].

Any molecule can be classified by its symmetry elements (i.e. axes and planes). It is then possible to assign the molecule to a group called a point group which has these same elements. This can then be used to predict which bands are infrared and which are Raman active. To do this it is necessary to work out the symmetry elements in the molecule. The main symmetry elements we need to recognize are the following:

E – The identity element. This takes the molecule back into the same position it started from, i.e. a 360° rotation for every part of the molecule does this.

C_n – An axis of symmetry in which the molecule is rotated about a molecular axis. n defines how often the molecule requires to be rotated to arrive back at the starting point. In the nitrate ion shown in Figure 3.5, one possible axis is

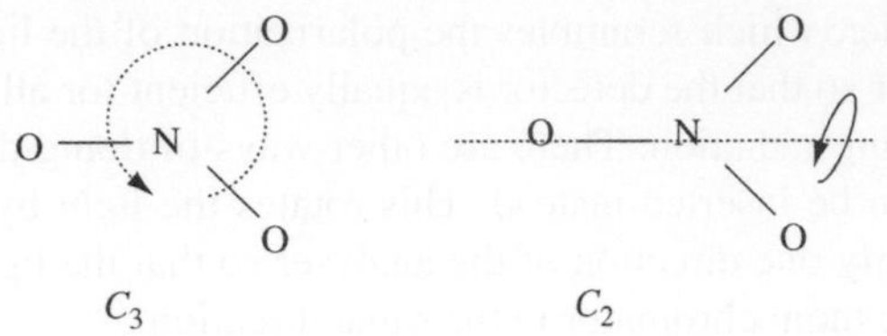

Figure 3.5. C_3 and C_2 axes in the nitrate ion.

the one pointing straight out of the plane of the paper. If the molecule is rotated about it, each oxygen will require to be rotated three times to arrive back at the starting point. This is a C_3 axis. There may well be a number of axes in a molecule. For example, in the nitrate ion, three C_2 axes also exist. They lie along the NO bonds and rotating the molecule about them would require two rotations to take the molecule back to its starting point. The axis with the highest value of n, for the nitrate C_3, is known as the principle axis of the molecule.

σ_h – A plane of symmetry in which the plane is perpendicular to the principle axis of the molecule.

σ_v – A plane of symmetry in which the plane is parallel to the principle axis of the molecule.

i – A centre of inversion in which every point inverted through the centre arrives at an identical point on the other side.

S_n – An axis which combines a rotation and an inversion.

The nitrate ion has one σ_h plane and three σ_v planes. These symmetry elements define a particular type of molecule.

All molecules with the same set of symmetry elements are said to belong to the same point group. To assign a molecule to its point group, the symmetry elements are first recognized and then analysed according to a set of rules. Usually symmetry elements are not analysed to assign a molecule to a particularly high symmetry point group such as the cubic point group, the octahedral point group and the tetrahedral point group. These can usually be recognized immediately. The questions we ask to make an assignment are set out in order below:

1. What is the principle axis of the molecule?
2. Is there a set of n C_2 axes at right angles to it? If the answer is no, carry on with the questions below. If the answer is yes, go to question 6.
3. Is there a plane perpendicular to the principle axis? If so, this is a σ_h plane. A molecule which has a C_n principle axis and a σ_h plane can be assigned to the point group C_{nh}.
4. If there is no σ_h plane, are there planes of symmetry parallel to the principle axis? There should be as many planes of symmetry as the n value. If this is the case, the point group is assigned as C_{nv}.

5. If there are no planes, the point group is assigned as C_n.
6. If the molecule has a principle axis and a set of n C_2 axes at right angles to it, is there a plane of symmetry perpendicular to the principle axis (i.e. a σ_h plane)? If this is the case, this molecule belongs to the D_{nh} point group.
7. If there is no σ_h plane of symmetry, is there a set of n σ_v planes parallel to the principle axis? If the answer to this question is yes, then the point group is D_{nd}.
8. If there are no planes of symmetry, the molecule will belong to the D_n point group.

Other molecules will belong to lower symmetry point groups. For example, for some molecules there is an S_n axis or a σ_v plane or no symmetry element at all. These can be recognized and a point group assigned by inspection.

Having assigned the molecule to a point group, group theory can be used to predict whether or not a band will be Raman or infrared active. It is particularly important to note that symmetry considerations allow us to determine whether or not a band is allowed in a Raman or infrared spectrum. This does not tell us how strong it will be; this would require a calculation.

There is a group theory table for all point groups which defines the symmetry behaviour of every vibration of a molecule belonging to that point group. Below, we reproduce the C_{2v} point group table which would be the correct point group for a single molecule of water.

C_{2v}	E	C_2	$\sigma_v(xz)$	$\sigma'_v(yz)$		
A_1	1	1	1	1	z	x^2, y^2, z^2
A_2	1	1	−1	−1	R_z	xy
B_1	1	−1	1	−1	x, R_y	xz
B_2	1	−1	−1	1	y, R_x	yz

In the table, the symmetry elements are shown across the top. The first column contains a series of letters and numbers. The first one we see is A_1. This is a way of describing a vibration, or for that matter an electronic function. It describes what happens to the vibration with each symmetry element of the molecule. These symbols are called irreducible representations and the top line always contains the one which refers to the most symmetrical vibration in terms of its behaviour when it is rotated or reflected by the symmetry operations. In higher symmetry point groups where there is a centre of symmetry, there would also be a g or a u subscript. For example, the most symmetric representation in the D_{4h} point group to which the molecule $AuCl_4^-$ belongs is A_{1g}. The vibration to which this corresponds is the left-hand vibration in Figure 3.2. There are four possible letters, A, B, E and T. A and B mean that the vibration is singly degenerate. E means it is doubly degenerate and T means it is triply degenerate. In the C_{2v} point group all vibrations are singly degenerate. A is more symmetric than B. Across the line

(a)   (b)

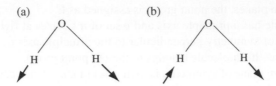

Figure 3.6. Two vibrations of water.

from the symbols representing the irreducible representations, there are a series
of numbers for each. The numbers are either 1 or −1 and 1 is more symmetric
than −1. For example, in the table, an A_1 irreducible representation gives the
value of 1 for every symmetry element.

Figure 3.6 shows two vibrations of water. By looking at the shape of the molecule it is possible to assign it to the point group C_{2v} using the methods given
above. For the vibration in (a), when the molecule is rotated about the C_2 axis, the
direction of the arrow representing a vibration does not change. This is the highest symmetry and is denoted as 1. In addition, the direction does not change when
the arrow is reflected by either of the planes of symmetry (the plane of the paper
and one perpendicular to it which bisects the oxygen). Therefore, the vibration
(a) is assigned to the highest symmetry irreducible representation of the C_{2v} point
group (A_1). In vibration (b) the sign of the arrow is reversed for C_2 and one plane.
When this happens this is given the number −1. Thus, vibration (b) belongs to a
lower symmetry representation. The actual label depends on which plane of symmetry is considered first in the table. It is conventionally given as the irreducible
representation B_1.

By this method we can assign a vibration to a particular irreducible representation
in a particular point group using the questions set out above. For more complex molecules there is a procedure to follow to do this and this is explained in books on the
subject [1, 4].

The main advantage of this assignment is that the tables also contain information
that enables us to work out whether the vibrations will be allowed by symmetry
or not. For infrared, this is done by multiplying the irreducible representation of
the vibration by the irreducible representation of x, y or z which is given in the end
column of the point group table in most, but not all, layouts. These correspond to
three Cartesian coordinates of the molecule and are the irreducible representations of
the dipole operator. If this result contains the totally symmetric representation (the
highest symmetry representation in a particular point group A_1 in the point group C_{2v}
but A_{1g} in the point group D_{4h}) then the vibration is allowed. The reason this works
is that a vibration can be allowed only if the product of the irreproducible representations of the ground state, the operator, and the excited state is totally symmetric
(i.e. the product of the irreducible representations of the components of the triple

integral discussed in Section 3.5). Since the ground state is always totally symmetric, it turns out that we only need to multiply the other two. A similar approach is adopted for Raman scattering but as discussed above the equation is more complex. In this case we look for the more complex quadratic functions x^2, y^2, z^2, xy, $x^2 - y^2$, etc. in the table and these are multiplied by the symmetry representation of the vibration. For simple point groups with nondegenerate representations, the rules for multiplying irreducible representations are $A \times A = A$, $B \times B = A$, $A \times B = B$, $1 \times 1 = 1$, $2 \times 2 = 1$ and $1 \times 2 = 2$.

It is unlikely that many readers will carry out an in-depth analysis of this type and in the interests of balance, the reader who requires more information is referred to a group theory book for a fuller explanation [3]. However, the symbols are often used in spectroscopy and the irreducible representations can be used to show if a band is allowed. All readers with serious interests in spectroscopy need to know what they mean.

3.10   LATTICE MODES

One type of vibration which has not been considered so far are vibrations created in solid samples by radiation interacting with a lattice rather than a molecule. For example, sodium chloride and silicon give Raman spectra but there is no definite molecule to give the scattering. The direction of the displacements with lattice modes is defined with reference to the incoming radiation. A longitudinal optic mode (LO) will travel along the direction of the incoming radiation and neighbouring ions will move along that direction but in opposite directions to each other. A longitudinal acoustic mode (LA) will travel along the same direction but the neighbouring entities will move in the same direction as each other. The transverse optic mode (TO), which is doubly degenerate, spreads out at right angles to the propagation direction and again neighbouring entities move in opposite directions to each other. The direction of the transverse acoustic mode (TA) is similar but neighbouring entities move in the same direction. All these modes are of low frequency but the acoustic modes are at lower frequencies than the optic modes which are usually the modes studied by Raman spectroscopy although spectra can be obtained from acoustic modes in some materials as can combination modes including them. The nature of the TO and TA modes is illustrated in Figure 3.7 for a simple model, a single line of sodium chloride atoms. These vibrations spread throughout the lattice and there are many different possibilities giving many individual vibrations with slightly different frequencies. Thus, what is observed experimentally for each mode is a broad band consisting of a very large number of individual vibrations. Bands of this type can be very diagnostic of solid-state arrangements in a wide variety of solids including carbon and silicon as described briefly in Chapter 6.

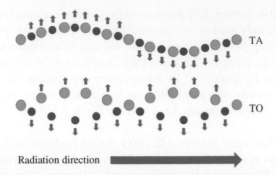

Radiation direction →

Figure 3.7. Lattice modes for a single line of sodium chloride atoms showing the greater charge separation in a transverse optical mode (TO) compared to the transverse acoustic mode (TA). Positively charged sodium ions are red and negatively charged chloride ions are green.

3.11   SUMMARY

The purpose of this chapter was to add depth to the basic methods needed to interpret a spectrum. The mathematics in this section can become quite complex and consequently salient equations are presented and explained. This complexity is not surprising since the equations have to describe the molecule in a distorted state at the instant that there is an interaction between the laser radiation and the molecule. However, some of the conclusions are quite simple and the analysis given above provides an insight into the background theory for Raman scattering, which helps with problems and more detailed interpretation. It also gives insight into the way computer predictions discussed in Chapter 2 can be obtained and paves the way to simplify the next chapter on resonance Raman scattering. Symmetry labels are commonly used in the literature, and at a basic level, the spectroscopist needs to understand their meaning to aid comprehension of many articles. The use of symmetry also improves understanding of the nature of vibrations and gives insight into the science that underlies the selection rules. Scattering theory is essential to understand the Raman process in depth and explain such features as the weakness of overtones and resonance. However, much of the rest of this book after Chapter 4 can be understood with only a minimum appreciation of the contents of this chapter.

REFERENCES

1. Long, D. (1977). *The Raman Effect: A Unified Treatment of the Theory of Raman Scattering by Molecules*. Wiley.
2. Ferraro, J.R. and Nakamoto, K. (1994). *Introductory Raman Spectroscopy*. San Diego: Academic Press.
3. Long, D.A. (2002). *The Raman Effect*. Chichester: Wiley.

4. Cotton, F.A. (1990). *Chemical Applications of Group Theory*. Wiley Interscience.
5. Clark, R.J.H. and Dines, T.J. (1986). *Angew. Chem. Int. Ed. Engl.* **25**: 131.
6. Clark, R.J.H. and Dines, T.J. (1982). *Mol. Phys.* **45**: 1153.
7. Rousseau, D.L., Friedman, J.M., and Williams, P.F. (1979). *Top. Curr. Phys.* **2**: 203.
8. Mie, G. (1908). *Ann. Phys.* **330**: 377.

Chapter 4

Resonance Raman Scattering

4.1 INTRODUCTION

Resonance Raman scattering occurs when the excitation frequency used is the same as that of a suitable absorption band. In the early days, many spectroscopists preferred to avoid coloured compounds. After all, if a powerful beam of visible radiation is used to excite a molecule which is coloured, the light is liable to adsorb into the sample. This can cause sample decomposition through photodecomposition or heating and may excite strong fluorescence preventing the detection of the Raman scattering. However, many coloured species do not fluoresce efficiently and, in resonance conditions, scattering enhancements of up to 10^6 have been observed and they are quite often of the order of 10^3 or 10^4. This makes resonance Raman spectroscopy a sensitive technique and, since only the chromophore gives the increased scattering, it is more selective. In contrast to normal Raman scattering, information on the electronic structure of a molecule can be obtained. It can be used to selectively identify a particular component in a mixture, for example, to study *in situ*, a coloured species in a reaction, dyes or pigments in inks, colours in paintings or in archaeology, proteins containing a chromophore such as a heme and labelled DNA and antibodies. There are more general implications for the interpretation of normal Raman spectra. The scattering does not follow the fourth power law and the more efficient scattering means that low concentrations of impurities can provide strong bands in the spectrum making it is easy to misinterpret them as arising from a bulk species. The theory is well described in a number of reviews (see Refs. [1, 2]). Here, we will not carry out a rigorous mathematical treatment but concentrate on understanding the differences between resonance Raman scattering and Raman scattering and the implications this has for the practical spectroscopist.

Modern Raman Spectroscopy: A Practical Approach, Second Edition. Ewen Smith and Geoffrey Dent.
© 2019 John Wiley & Sons Ltd. Published 2019 by John Wiley & Sons Ltd.

4.2 THE BASIC PROCESS

To obtain resonance Raman scattering, an excitation wavelength is selected which is close to or the same as that of an electronic transition. Ideally, a tunable laser should be used to provide a frequency to maximise scattering but it is often more practicable to use a laser available in the laboratory, which has a frequency close to the resonance frequency. In resonance, the molecule may absorb or scatter each photon which interacts with it. The relative efficiency of absorption and scattering is a property of the molecule and some molecules are much more efficient scatterers than others. As we know from Chapter 3, the scattering process is faster than the absorption process, with scattering occurring before the nuclei reach equilibrium positions, in this case for an actual excited state. Thus, the processes of resonance Raman scattering and absorption are differentiated by time. In addition, after absorption, whether a molecule loses the energy gained through non-radiative processes or by fluorescence is dependent on the nature of the molecule making it easier to record the scattering from some chromophores than others.

In absorption spectroscopy, the radiation irradiating the sample is often polychromatic and transitions to many excited states will occur. Often, the most intense transition is to one of the higher vibronic levels. The round shape of most absorption bands is usually due to contributions from transitions to many vibronic states and to the presence of hot bands arising from electrons present in excited levels of the ground state. However, as we shall discuss later in this chapter, the theory of Raman scattering predicts two different types of resonance enhancement, A type and B type. B-type enhancement is common in larger molecules and comes only from the first two vibronic levels of the excited state in resonance. A-term does not have the same restriction but the scattering efficiency from each vibronic level is not the same as the efficiency of absorption in electronic spectroscopy. As a result, it is normally not the case that the maximum absorbance of an electronic transition is at the energy at which the greatest resonance Raman scattering will be obtained. Usually it is at a position on the lower energy side of the absorption band.

There are also practical reasons for using excitation on the low-energy side of the absorption maximum as described later in this chapter. Briefly, absorption of the exciting radiation will limit the penetration depth of the excitation beam and crucially, the weaker scattered radiation will be absorbed by the medium – a process called self-absorption. Working just on the low-energy side of the absorption band will reduce these effects and can improve the sensitivity. Further, using exciting radiation on the low-energy side of the transition also makes fluorescence a less-efficient process.

4.3 KEY DIFFERENCES BETWEEN RESONANCE AND NORMAL RAMAN SCATTERING

It may be intuitively obvious that if the excitation radiation has the same energy as an excited vibronic state, there can be a stronger interaction between the excitation and the molecule causing larger changes in polarisation and more intense scattering. However, the resonant process is more complex than that giving selective enhancement of some bands in the chromophore and arising from two different processes which have different influences on the spectrum.

4.3.1 Intensity Increase

The increase in intensity from resonance enhancement can be understood by studying the KHD equation analysed in Chapter 3 (Eq. 3.5). Here, we will explain how the mathematics predicts the changes which occur in resonance with no further development of the equations introduced in Chapter 3. Fuller accounts of the mathematics and more in-depth references can be found in Refs. [1, 2]. Consider the denominator of the first term in Eq. (3.5). The resonance condition for Stokes scattering is met when the energy difference between the lowest vibrational state of the ground electronic state G and the resonant vibronic state I (or ω_{GI} in Eq. 3.5) is of the same energy as the exciting light ω_L. This would mean that the denominator of the first term reduces to $i\Gamma_I$, which is a small correction factor required to take account of the lifetime of the excited state. Thus, under resonant conditions, the denominator is small and this will lead to the term for the one state to which this applies becoming very large, increasing polarisability and giving very much greater Raman scattering. Since scattering from this state will predominate, scattering from other states can be ignored and the summation in Eq. (3.5) can be removed as shown in Eq. (4.1) with the second term of Eq. (3.5) dropped as explained in Chapter 3.

$$\left(\alpha_{\rho\sigma}\right)_{GF} = k\frac{\left\langle F|r_\rho|I\right\rangle\left\langle I|r_\sigma|G\right\rangle}{\omega_{GI} - \omega_L - i\Gamma_I} \tag{4.1}$$

One immediate consequence of the selection of one vibronic state is that the actual intensity observed will depend on the electronic nature of the specific excited state I selected by the resonance so that some vibrations will be more enhanced than others. Predicting intensities for specific vibrations requires calculation, but there are some simple things which can provide some idea. Where symmetry plays a part, as in heme systems, only some vibrations will be allowed so this can be used as a starting point to predict intensities. Assuming no symmetry or that symmetry selection rules have already been taken into account, intense bands are more likely to arise from transitions for which the vibration takes a molecule from the geometry of the electronic ground state to the electronic geometry of the resonant excited state. The reason for this is that the scattering process is faster than the vibrational process so only a small geometry change can occur during a scattering event. Thus, over the time of a vibration, overlap between the ground and excited electronic states at every possible geometry maximises the number of scattering events which will occur.

A good example of this is in the resonance Raman spectrum of 5,5′-dithiobis-2-nitrobenzoic acid (ESSE) and an ion (ES⁻), which arises from it (Figure 4.1). This is a standard reagent used in electronic absorption spectroscopy to analyse for thiols in

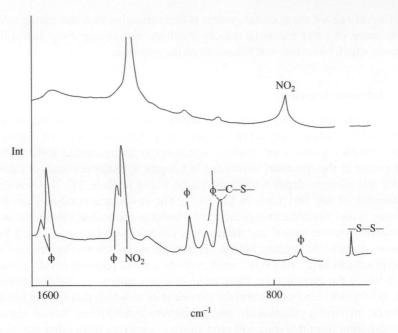

Figure 4.1. The spectrum of Ellman's reagent at 10^{-2} M (foot) and of the anion ES$^-$ at 10^{-5} M (top).

molecules since it reacts to provide a coloured ion as shown below. The compound is often called Ellman's reagent, hence ESSE.

$$ESSE + RSH = ESSR + ESH$$

$$PH\ 7:4$$

$$ESH = ES^- + H^+$$

ESSE has an absorption band at 325 nm. However, the coloured ion ES$^-$ formed at neutral pH has an absorption band at 410 nm. When a laser with an excitation wavelength of 457.9 nm (this was the closest wavelength to 410 nm available in the laboratory at the time) was used to excite ESSE, a typical Raman spectrum was obtained with bands assigned as due to vibrations of the phenyl rings (ϕ), the nitro group and the S—S bond. This gives an example of normal or perhaps pre-resonant spectra with most vibrations that would be expected to occur in the spectral region studied being present in the spectrum. The Raman spectrum from ES$^-$ is near to resonance and is much stronger. The spectrum of ESSE was obtained from a 10^{-2} M solution, but to obtain approximately the same scattering intensity, the spectrum of ES$^-$ was recorded at 10^{-5} M. Because ES$^-$ is near to resonance, the bands which are strongly resonantly enhanced dominate the spectrum clearly showing the selectivity of the resonance effect. The intense bands are assigned to NO_2 group displacements with the most intense one being the symmetric stretch. This suggests that the excited state has an electronic geometry which is elongated along the N—O bonds thus giving good overlap at any point during the vibration between the ground and excited states.

4.3.2 Franck Condon and Herzberg Teller Scattering

It is an oversimplification to consider only one state in resonance since other states close in energy will have quite small denominators and could contribute significantly but the general conclusions reached below are not affected so only one state is considered. Since the effect arises from one state, the closure theorem, which states that the sum of the products of all the vibrational states of a molecule is zero, is no longer valid. This theorem was the reason that the A-term in Eq. (3.10) did not predict any Raman scattering and since it is no longer valid, the A-term as well as the B-term can give resonance Raman scattering. This leads to two forms of resonance Raman scattering, which have quite different properties. These two terms are often called Franck Condon and Herzberg Teller scattering, respectively.

In A-term or Franck Condon scattering, the excitation which causes the scattering simply couples the ground state and the excited state. In Eq. (3.10), where the wave functions of the states have been separated into the electronic and vibrational parts using the Born Oppenheimer approximation, the electronic term (M) in A-term scattering is M^2 as opposed to $M \times M'$ in B-term where M' is the derivative of M. This would suggest that A-term scattering can be more intense than B-term scattering. However, this is only one factor. Two specific states are involved in resonance scattering so the size of the numerator depends on the two wave functions involved.

Allowed electronic transitions give intense scattering and if the molecule has symmetry, symmetry selection rules will apply. In addition, strong A-term resonance enhancement occurs when there is a difference in the nuclear geometry between the ground and excited states as in the case of the NO_2 group in ES$^-$ above. This latter condition means that small molecules usually have strong A-term enhancement. However, it does occur in large molecules even if there are only small displacements on each bond as in a phenyl ring since the sum of these displacements can be appreciable.

The coordinate operator R_ε in the expression for B-term or Herzberg Teller scattering in Eq. (3.10) can mix states from the starting geometry to make the excited state. For example, if there are two $\pi \rightarrow \pi^*$ transitions reasonably close together in the visible region, as is the case with porphyrins, then the coordinate operator can mix them. This allows a resonant vibronic state of a forbidden electronic transition to borrow intensity from an allowed vibronic state and give resonance scattering. This is important in porphyrins since the lower energy $\pi \rightarrow \pi^*$ transition of porphyrins, which gives rise to the Q bands, is forbidden and weak in the absorption spectrum, but B-term scattering from this transition is appreciable if less than the A-term scattering, which arises from the allowed Soret band. Further, because of the high symmetry of the porphyrin core, symmetry selection rules apply. For the A-term enhancement from the Soret band, the rules are as before that the most symmetric vibrations are the most enhanced (A_{1g} vibrations). The lower energy Q bands arise from a forbidden transition but borrow intensity from the Soret band, enabling B-term enhancement. However, the extra term required to mix the electronic transitions in B-term makes the less-symmetric gerade B_{1g} and B_{2g} vibrations the most intense [1–4].

Figure 4.2 shows the absorption spectrum and Raman scattering with 406 nm excitation from a protein with a heme group. The allowed B or Soret transition is much more intense than the forbidden Q bands, which borrow intensity from it. If the porphyrin ring is regarded as flat and peripheral groups which make up the heme such as the vinyl groups are ignored, it belongs to the D_{4h} point group. 406 nm excitation will produce resonance with the allowed Soret band, and the totally symmetric A_{1g} modes are the most enhanced indicating A-term scattering. This spectrum is remarkable not only because it shows the clear enhancement of A_{1g} modes but because no bands from the rest of the protein can be observed due to the resonant enhancement of the heme. The displacements which produce three of the most intense peaks are shown in Figure 4.2. The absorption spectrum for the Q bands shows some vibronic structure which in many compounds is not observed at room temperature because of the large number of overlapping transitions which occur and combine to make up the spectrum. This enables resonance Raman spectra to be taken both for the fundamental 0–0 and the 0–1 electronic transitions. As stated above, B_{1g} and B_{2g} transitions predominate to give a very rich and complex spectrum, which can be used to probe the nature of the excited state. The spectra and an analysis are given in Ref. [1].

The intense A_{1g} bands, which dominate with 406 nm excitation, are used as markers for the oxidation state and spin state of the ion. The oxidation state marker, ν_4, shifts more in frequency than other bands when the oxidation state of the iron ion in the centre of the ring is changed. One of the largest displacements is on the four nitrogens attached to the central metal ion. They move symmetrically inwards and outwards in phase to alter the size of the hole in the centre. Thus, the size of the metal ion which fills this hole and which changes as the oxidation state changes would be expected to have a significant effect on the frequency of this vibration. The reason why ν_3 and ν_{10} act as spin-state markers of the iron is more subtle. Their common feature is that they have significant displacements on the inner ring system of the porphyrin.

A key difference between A- and B-term enhancement is that because of the coordinate operator, B-term enhancement occurs only from the zero and first vibronic states of the excited state in resonance. There is no restriction on the excited vibronic states which can give A-term enhancement. Initially it was thought that larger molecules such as those containing aromatic ring systems which have smaller nuclear displacements per bond would give only B-term enhancement but it is now clear that the sum of the displacements is significant and both B-term and A-term enhancement can occur in a larger molecule.

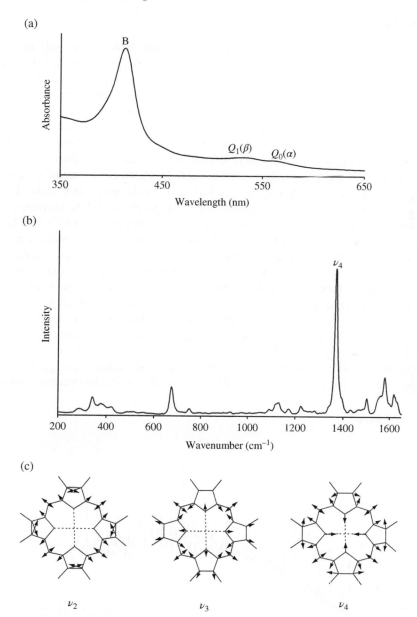

Figure 4.2. Spectra and displacements for the heme chromophore in a protein. (a) Electronic spectrum showing the allowed B or Soret band and the forbidden Q bands. (b) Raman spectrum using 406 nm excitation, which is in the Soret band region and (c) three A_{1g} vibrations of the porphyrin ring.

4.3.3   Overtones

As described in Chapter 3, in normal Raman scattering, which arises only from the B-term, the coordinate operator R_e allows a transition only if there is one vibrational quantum number difference between the ground and the excited states (see Eq. 3.10). That is, no overtones should appear. However, in A-term scattering there is no coordinate operator in the numerator. As a result, where resonance arises from A-term enhancement, overtones will be allowed and there is no reason why they should not be intense. The Raman spectra of iodine are a classic example of A-term resonance enhancement from a small molecule. Excitation of I_2 vapour obtained by heating some iodine in a gas cell produces a remarkable spectrum consisting of a series of sharp lines, which are very intense (see Ref. [1]). As discussed in Chapter 1, one vibration is expected for a diatomic molecule in the gas phase and the frequency will correspond approximately to the energy we would expect from Hooke's law. However, what is obtained is a long series of bands at regular intervals and the energy difference between them is approximately one quantum of the energy expected for the single fundamental vibration. A similar experiment is easily performed either by using a solution of iodine or an iodine crystal. Figure 4.3 shows the spectrum of solid iodine recorded from an iodine crystal. Again, there is a regular pattern of bands.

We are able to learn more about the nature of iodine from this spectrum than we would from conventional Raman scattering. This is clearly A-term scattering in which there is no selection rule to forbid the overtones occurring. With the harmonic approach, we would expect the overtones to be equally separated in energy. However, as discussed in Chapter 3, with the true Morse curve, the separation decreases towards

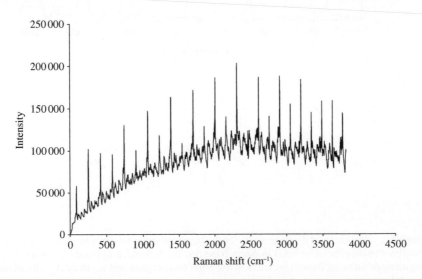

Figure 4.3.   Spectrum of iodine in the solid state obtained using a Raman microscope and 514.5 nm radiation.

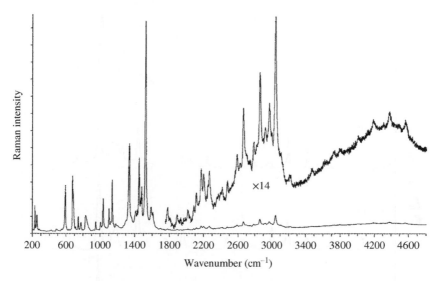

Raman intensity

Wavenumber (cm⁻¹)

×14

Figure 4.4. Spectrum with overtones of copper phthalocyanine taken with an excitation frequency of 514 nm. Source: Taken from Tackley, D.R., Dent, G., and Smith, W.E. (2001). *Phys. Chem. Chem. Phys.* **3**: 1419 [5] with permission from The Royal Society of Chemistry.

higher energies as a result of the nonharmonic nature of the curve. Thus, by studying the changes in separation, it is possible to calculate the shape of the Morse curve. This is one example of Raman scattering providing electronic information.

Larger molecules also produce overtones in resonance conditions. Usually they are much weaker and decay quickly. Figure 4.4 shows this for copper phthalocyanine. These spectra are quite complex to analyse. They contain not only overtone bands which consist of more than one quanta of the same vibration but also combination bands consisting of one quanta from two different vibrations. The fundamental bands occur up to about $1600\,cm^{-1}$ and are followed mainly by first overtones up to $3200\,cm^{-1}$ and by some weaker second overtones beyond that. The most intense first overtone corresponds to two quanta of the most intense fundamental band which is assigned as ν_3 and has A_{1g} symmetry in keeping with A-term enhancement.

4.3.4 Wavelength Dependence

Normal Raman scattering is dependent on the fourth power of the frequency and the absolute intensity will increase according to that law as the frequency of the excitation is increased. However, the relative intensities of the bands in spectra obtained with different excitation frequencies will be the same. In resonance Raman scattering, the intensity of the scattering from vibrations, which are resonantly enhanced, will increase as resonant conditions are approached so that the relative intensities of bands will change. Thus, the relative intensity of the bands varies with frequency dependent

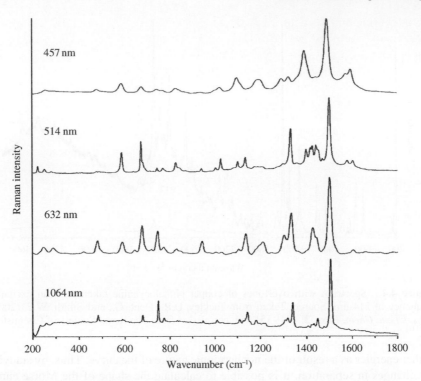

Figure 4.5. The spectrum of copper phthalocyanine taken with four different excitation frequencies. Source: Taken from Tackley, D.R., Dent, G., and Smith, W.E. (2001). *Phys. Chem. Chem. Phys.* **3**: 1419 [5] with permission from The Royal Society of Chemistry.

on the nature of the electronic structure. The copper phthalocyanine spectra shown in Figure 4.5 are an illustration of that. The spectrum taken with 1064 nm excitation may be close to a normal Raman spectrum since phthalocyanines have no obvious electronic transitions in that region. Using that spectrum as a reference, it is clear that bands rise and fall in intensity across the visible region depending on the enhancement of each specific vibration caused by the proximity of the two $\pi–\pi^*$ transitions.

The 1064 cm^{-1} spectrum is described only cautiously as close to a normal Raman spectrum despite this excitation frequency being well away from the frequency of any electronic transition. The question arises as to how far from the resonance frequency does the excitation frequency have to be to show no resonance. It is worthwhile to look at the KHD expression once again. As previously discussed, in the resonance condition the denominator is very small. However, as the difference between the laser excitation frequency and the transition frequency increases, the resonance enhancement will drop off quite quickly. If arbitrary units are used so that the numerator can be given the value 1, then 10 wavenumbers away from resonance, and ignoring $i\Gamma_r$, the enhancement will be down to about a tenth and at 100 wavenumbers about a hundredth, etc. (This is an oversimplification since the numerator will vary with frequency as other states come into resonance but we can neglect that to make a qualitative conclusion.) It is clear that it is necessary to be quite close to resonance to get the greatest enhancement and an enhancement of two or three orders of magnitude will have little effect on the spectrum a few hundred wavenumbers away from the resonant condition. Perhaps of more importance and not widely recognised by many Raman spectroscopists is what happens with larger enhancements well away from the resonance condition. Supposing, for example, the total enhancement obtained was 10^4, then only 10 000 wavenumbers away from resonance, would the enhancement be down to 1. Thus, there is a very long tail on the frequency dependence of the resonance enhancement process and it could be that when a chromophore in the red region of the spectrum is present and infrared excitation is used, small enhancement factors could be obtained. This may seem trivial compared to the enhancements that can be obtained close to resonance but it does mean that bands from a chromophore could easily be more intense than those from the rest of a molecule. Further, if a minor component of a mixture is a strong resonance Raman scatterer, it could mean that it may be picked out selectively.

Scattering taken close to but not at the resonance condition is usually called pre-resonance scattering and will give appreciable enhancement with a strong scatterer. If the excitation frequency is lower than the frequency range of the absorption band, fluorescence and absorption will not occur from the resonant state making recording of the Raman spectrum simpler. We will give an example later (Chapter 6) describing Raman scattering obtained from ink jet dyes with infrared excitation.

4.3.5 Electronic Information

Two ways in which electronic information can be obtained have already been described, namely in the use of overtones with iodine to define the shape of the Morse curve and the electronic structure which can be deduced from the enhancement of the NO stretch in the ES^- ion from Ellman's reagent. However, a more detailed analysis is possible and is now much easier with the availability of good tunable lasers. Since the absolute and relative intensities of the bands in a resonant spectrum are dependent on how close the excitation frequency is to the resonant frequency and on the nature of the electronic states, it can be useful to plot the intensity of a band against the frequency of the excitation used. This is called a resonance excitation profile (REP) and to construct one, a large number of spectra taken at closely spaced frequencies are required. In the simplest REP, the maximum intensity of the band produced will be the point at which resonance occurs. If the vibration for which the intensity is plotted is resonant with more than one vibronic level, more than one peak should be seen in the profile. Also, as discussed in the sections above, different vibrations couple differently to the electronic structure so the profiles will vary from one vibration to another. The information obtained provides unique and extremely

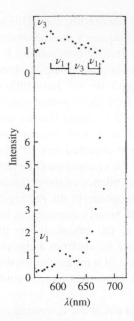

Figure 4.6. Resonant excitation profiles for two copper phthalocyanine vibrations ν_3 and ν_7. They were obtained by taking spectra at closely spaced wavelengths and measuring the intensity of the selected vibrations. Source: Taken from Bovill, A.J., McConnell, A.A., Nimmo, J.A., and Smith, W.E. (1986). *J. Phys. Chem.* **90**: 569 [6]. Copyright the American Chemical Society.

valuable information to probe in depth the electronic and vibrational structure of a particular molecule. An example of a resonant excitation profile for two different vibrations is shown for copper phthalocyanine in Figure 4.6. One, ν_3, is at a relatively high frequency and the structure on the profile indicates that a number of vibronic states are involved, suggesting A-term enhancement. The band is still rising at the higher energy of the range studied suggesting more vibronic states give enhancement. The lower frequency vibration, ν_7, is more typical of B-term enhancement with one strong peak from the $\upsilon = 0$ states of the ground and excited states and weaker coupling between $\upsilon = 0$ and $\upsilon = 1$ states only. A weaker peak can also be seen on the low-energy side of the main band. This is because the excited state of the phthalocyanine is split by a dynamic Jahn-Teller effect and the extra weak band arises from a second electronic state.

The above description provides only a qualitative understanding of the nature of resonance Raman scattering and some issues have been presented without real proof. A more in-depth review can be obtained in references 1 and 2. However, the salient points have been covered to enable practical use of resonance Raman scattering and interpret the spectra. Table 4.1 gives the main similarities and differences between Raman scattering and resonance Raman scattering.

Table 4.1. Main differences between Raman scattering and resonance Raman scattering.

Raman scattering	Resonance Raman scattering
B-term effective	A- and B-term effective
No overtones	Overtones common
More modes observed in the spectrum	Some modes selectively enhanced
No electronic information	Electronic information present
Weaker scattering	Stronger scattering

4.4 PRACTICAL ASPECTS

The selection of the best frequency to give maximum enhancement can be difficult and often easiest to establish experimentally with a tunable laser. Figure 4.7 shows a simplified picture for one vibration with only one of the many possible transitions which will contribute to the absorption process and one possible resonance Raman process shown. It assumes there are no symmetry selection rules. The excited state is usually slightly broader than the ground state and with a minimum at a slightly greater internuclear separation. The most intense peak contributing to the absorption spectrum will be where the maximum overlap occurs between the ground and excited electronic states. Usually this is a transition from the middle of the ν_0 ground state to near the side of a higher vibronic state of the excited state, but there are many transitions. In contrast, in Raman spectroscopy, if only B-type enhancement is strong, the first two vibronic levels will provide the scattering. Only one process, $\nu_0 - \nu_0$, is shown. In the absorption spectrum, hot bands due to transitions from populated excited vibrational levels in the ground electronic state will be present on the low-energy side of the band. As a result, the excitation frequency required to be in resonance with the ν_0 and ν_1 levels is normally on the low-energy side of the absorption band and slightly into it but not at the peak. Maximum A-term enhancement is likely to be more broadly spread across the band but is not likely to be at a maximum at the same frequency as the maximum in the absorption spectrum. Thus, it is often better to excite into the low-energy side of the absorption band and this has the advantage that it reduces problems discussed below with absorption of the excitation and scattered radiation.

Since in resonance Raman scattering the excitation energy corresponds to that of an absorption process, absorption can readily cause sample decomposition and/ or fluorescence. Therefore, the spectroscopist must develop a strategy which minimises these effects. To assess the extent of the problem, one obvious thing to do is to observe the sample before and after exposure to the laser beam. Very often with resonance Raman scattering, sample damage can be clearly seen as a change in colour or a black spot. However, in some cases the damage is more subtle. For example, in proteins containing the heme group, radiation can alter the conformation of the heme and its bonding in the protein without destroying it. One simple check

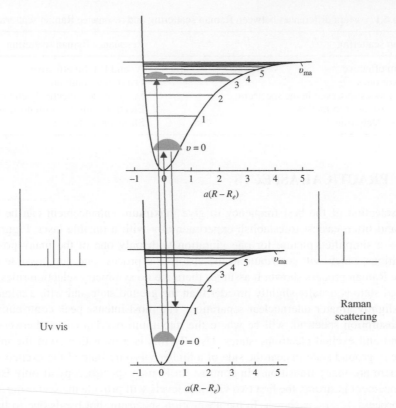

Figure 4.7. Morse curves for the ground and excited state and an illustration of the difference, other than time in the excited state, between absorption and resonance Raman scattering. Only one of the many transitions which make up an absorbance spectrum is shown. With no special selection rules, the intensity of an absorption band depends on the maximum overlap between the ground and excited state wave functions. Commonly this is between the middle of the v $v = 0$ ground state and an off-centre part of a higher excited state as shown. Arrows for only the state giving the greatest absorption is shown for clarity (blue arrow). The spectrum on the left shows a contribution from a number of transitions but for many molecules, there are so many that a smooth broad peak results. With B-type resonance, since only transitions involving the first and second states ($v = 0$ and $v = 1$) are allowed, maximum enhancement will be with an excitation frequency on the low-energy side of the absorption peak. Only one vibration is shown (red arrows) since this diagram shows the ground and excited vibronic levels for one vibration. The weaker second peak represents an overtone. A-type enhancement is not restricted in this way and can occur from higher states but the selection rules are not the same as for absorption spectroscopy.

for damage is to run the absorption spectrum. However, in solution, there is often a very significant difference in sample volume interrogated by Raman scattering and absorption spectroscopy and this can lead to misleading conclusions. In most absorption spectrometers, the sample is placed in a 1 cm cuvette and the beam passes through a significant part of the sample volume whereas if Raman scattering is obtained from the same cuvette, it is usually obtained from a small focused volume in which there is a much higher power density. Damage, which occurs in this volume, will affect the Raman scattering, but it may not be sufficient to be observable in the absorption spectrum of the bulk of the sample. In a solid, reflectance absorption spectra are usually taken from a larger area and will be dominated in most cases by undamaged material.

One way to minimise photodegradation is to use a sampling method in which the sample passes through the laser beam but does not stay in the beam for the whole time of the analysis. Raman scattering is then obtained from the cumulated spectra from a large area or volume of the sample. For example, with solid samples, spinning disks can be used, as already mentioned in Chapter 2. In this technique, the sample is pressed into a disk, or compressed into a channel cut in a black support disk. The laser is then set to focus on the outer part of the disk or onto the channel and the disk spun. Scattering is collected from the point at which the beam is focused but the sample precesses through the beam limiting the exposure time of any one area. This also allows excited states to relax and heat to diffuse away from any one point before the disk completes a revolution and the same part of the sample is interrogated again. Alternatively, the optics can be set so that the beam is focused off centre and precesses over the sample in a circle. This is effective, although in some samples, the track where the sample has decomposed can clearly be seen. If this is present, parameters can be changed, for example, by using a lower laser power, shorter exposure time or increased spot size. The effect of decomposition can then be assessed to some extent by comparing the differences in the two spectra. A similar approach can be used with solution samples (Section 2.5). The devices usually use either a spinning sample container, such as an NMR tube with the beam tightly focused into the solution, or a small flow cell where the sample can either flow through or oscillate backward and forward under the laser beam.

The spectroscopist has another approach which may be effective. As discussed earlier in this chapter, the resonance contribution extends over a range of frequencies decreasing as the difference between the resonant frequency and the excitation frequency increases. Thus, by moving the frequency of excitation away from the resonance frequency to give pre-resonance, it is possible to avoid the worst effects caused by absorption of the excitation while still retaining a degree of enhancement. With pre-resonance excitation as discussed above, it should be borne in mind that the further away from resonance the spectrum is recorded, the more states will contribute and the more normal Raman scattering selection rules will apply. In addition, for molecules of high symmetry, as the frequency is scanned away from resonance, vibrations of different symmetry have different dependencies on frequency.

In a nonresonant sample, a visible laser beam can be focused tightly within the media and although there are refraction, reflection and heating effects which can cause problems in some samples as described in Chapter 2, the focused spot can be effectively imaged back onto the detector. However, focusing into a coloured solution to obtain resonance causes additional changes. The laser light is absorbed by the medium and the deeper it penetrates, the less intense is the light. Thus, the focused spot where the maximum power density should occur may be deep enough in the sample that there is relatively little power left. In addition, the absorbed energy can heat the environment causing more complex refraction (lensing) which causes the interrogation volume to change shape and decrease the efficiency of collection of the scattered radiation. Focusing close to the surface can help reduce these effects but brings complications as discussed below. Perhaps the most important problem is that the scattered radiation is much weaker than the exciting radiation and is absorbed as it travels back through the solution. This effect, self-absorption, can make resonance Raman scattering difficult to obtain from any depth despite the enhancement.

In solution, there is usually a concentration range in which resonance scattering is effective. If the sample is too concentrated, self-absorption of the scattered radiation will prevent effective collection of the Raman scattering. However, if the sample is too dilute, the Raman scattering will be too weak to detect. Thus, it is important while measuring solution resonance Raman scattering to recognise that if poor scattering is obtained, it may either be necessary to dilute the sample to allow penetration of the beam or to increase concentration to increase the number of chromophores. It is usually easier to find the right concentration by trial and error.

To minimise these effects, the beam should be focused close to the sample surface, but there is a limit as to how well this can be done. In a solid, specular reflection from the surface can occur if the beam is focused directly on the surface, and in solution, focusing too close to the glass wall at the front of the sample can cause intense reflection from the glass. The sudden appearance of more intense radiation usually means that the laser is focused on the glass rather than on the solution. This intense scattering can mislead the spectroscopist into recording spectra of the wall of the vessel rather than the solution. The ready availability of disposable plastic cuvettes has made this a more serious problem. When focusing through them, the spectrum of the plastic is not usually observed, but if the beam is focused on the wall of the cuvette, excellent spectra can be recorded from the polymer material which can easily be mistaken for the spectrum from the sample. If no container is used as, for example, for a drop on a microscope slide or a solution in a Petri dish, heating can produce currents and evaporation and can make it difficult to get stable signals over time.

4.5 SUMMARY

Effective resonance Raman scattering will not be obtained for all molecules which are coloured. Much depends on the relative efficiency of the scattering and fluorescence processes. Fluorescence can dominate the spectra in some cases making it very difficult to obtain the Raman scattering experimentally. Sample decomposition and difficulties with self-absorption are also problems. However, for many systems the scattering is strong and the fluorescence weak and there are strategies available to get round the worst interferences in others. These include the use of pre-resonance to avoid the worst problems with fluorescence and the use of spinning sample holders or flow cells to reduce photodegradation. There are key advantages in using resonance Raman scattering which make it worthwhile. It provides more intense spectra and consequently can be used to selectively pick out and positively identify a molecule in a matrix. Electronic information about a molecule can be obtained from the intensities of the bands found in resonance, from the energy separations in overtone progressions, and from the overtone patterns that can be obtained. The weak nature of ordinary Raman scattering from molecules such as proteins and, in particular, water, make it possible to examine resonance Raman scattering from chromophores directly in the presence of some other materials. This makes resonance Raman scattering a particularly useful form of Raman spectroscopy in some areas with either added labels or natural chromophores including UV chromophores and it will be the basis for some of the work described in later chapters.

REFERENCES

1. Clark, R.J.H. and Dines, T.J. (1982). *Mol. Phys.* **45**: 1153.
2. Rousseau, D.L., Friedman, J.M., and Williams, P.F. (1979). *Top. Curr. Phys.* **2**: 203.
3. Spiro, T.G. and Li, X.-Y. (1988). *Biological Applications of Raman Spectroscopy*, vol. 3 (ed. T.G. Spiro), 1. New York: Wiley.
4. Hu, S.Z., Smith, K.M., and Spiro, T.G. (1996). *J. Am. Chem. Soc.* **118**: 12638.
5. Tackley, D.R., Dent, G., and Smith, W.E. (2001). *Phys. Chem. Chem. Phys.* **3**: 1419.
6. Bovill, A.J., McConnell, A.A., and Nimmo, J.A. (1986). W. E. Smith. *J. Phys. Chem.* **90**: 569.

Chapter 5

Surface Enhanced Raman Scattering and Surface Enhanced Resonance Raman Scattering

5.1 INTRODUCTION

Surface enhanced Raman scattering (SERS) requires that the analyte is adsorbed onto a suitably roughened surface, usually of gold or silver, before the spectrum is recorded. The reasons SERS can be attractive are that it can provide greatly improved sensitivity along with selective identification of an analyte in a mixture without separation. Practically, it is quite simple to obtain a spectrum from a suitable analyte. However, SERS has different selection rules to Raman scattering making interpretation of the spectrum more challenging than for normal Raman scattering. Controlling a system to obtain reproducible and reliable results requires care. Despite this, with reported enhancement factors of 10^6–10^{15} for suitable analytes, there is a strong attraction for those wishing to use Raman spectroscopy for very sensitive and informative analytical procedures. In addition, the problem of understanding the nature of metal surfaces under water is important, for example, in the study of electrolytic processes or corrosion and, for the metals for which SERS is applicable, it can give useful *in situ* information.

SERS was initially observed in 1974 by Fleischman, Hendra and McQuillan [1]. They reported strong Raman scattering from pyridine adsorbed from an aqueous solution onto a silver electrode roughened by means of successive oxidation–reduction cycles. The authors attributed the effect to a large increase in the electrode surface area caused by the roughening process which enabled more pyridine molecules to be absorbed. However, Jeanmarie and Van Duyne [2] and Albrecht and Creighton [3] showed that the intensity was due to more than the increase in

Modern Raman Spectroscopy: A Practical Approach, Second Edition. Ewen Smith and Geoffrey Dent.
© 2019 John Wiley & Sons Ltd. Published 2019 by John Wiley & Sons Ltd.

surface area. They noted that the likely increase in intensity from the roughening of the surface would be less than a factor of 10, whereas the enhancement obtained was of the order of 10^6. A cell used by us to repeat the first experiment and the spectrum of pyridine taken with different voltages applied to the cell is shown in Figure 5.1. As can be seen, the magnitude of the surface enhancement and the relative intensities of the peaks change depending on the voltage applied.

In addition to providing a suitably roughened surface of an appropriate metal, it is necessary to consider the chemistry of the process. It is important that there is effective surface adsorption of the analyte and preferable that the surface remains stable for the time of the measurement. If these points are neglected poor, variable results may be obtained. For example, it is possible to make a rough surface of iron and get some surface enhancement, but usually iron has been found to be ineffective. This is due to the rapid formation of a thick surface oxide layer which acts as a barrier to contact between the metal and the analyte. It also may cause a loss of the surface roughness.

SERS experiments can be carried out in a wide range of environments. These can vary from an atmosphere-controlled or vacuum chamber containing small amounts of pure substrate and analyte to cuvettes open to the air and containing biological material as well as the analyte. Only a few authors fully describe the chemical precautions they take in creating a surface. For example, silver readily oxidizes in aqueous solution and bubbling oxygen or nitrogen gas through solutions during substrate formation has a considerable effect on the substrate formed. Particle size and shape are affected and in some circumstances even silver wires can be formed. Further, with complex systems such as biological samples, preferential absorption of components other than the desired analyte can prevent effective adsorption. Thus, before beginning a SERS experiment it is best to make sure that the adsorption process is at least partly understood. It is also necessary to choose the correct metal and roughen the surface appropriately.

Before continuing, it should be recognized that there are many ways in which increased Raman scattering compared to that of a reference sample can be measured from a particular surface and some papers refer to these as SERS. Different surfaces give different scattering intensities for many reasons including different roughness, different depth of penetration of the beam, different orientation of crystallites, and, with substrates, the number of molecules which may be adsorbed on a specific surface or the formation of a resonant species on adsorption. Thin films can trap light and increase scattering intensity from multiple internal reflections. Perhaps the closest to metal-based enhanced scattering is from particles larger than the wavelength of light where, on excitation, light is trapped inside the particle creating strong fields on the surface and intense scattering from molecules adsorbed on the surface. However, the SERS referred to in this book is that defined by the International Union of Pure and Applied Chemistry (IUPAC) being, the phenomenon by which the intensity of vibrational bands in the Raman spectra of molecules within a few nanometres of the

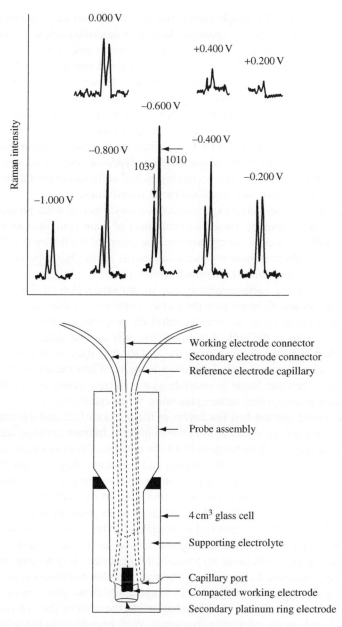

Figure 5.1. The spectra of pyridine at various potentials given as relative to SCE and a simple electrochemical cell of the type used to obtain the spectra. Peak positions are in cm⁻¹.

surface of microscopically rough metals, metal colloids and metal nanoparticles is increased by several orders of magnitude. As will be described below, the key feature theoretically is that the process involves surface plasmon assistance.

Silver and gold can be used to make particularly good substrates for SERS. Large enhancement factors can be obtained with visible or near-infrared (NIR) excitation from surfaces which are sufficiently stable to provide time-stable results. Other metals are also effective to varying degrees. Copper gives good enhancement but is more reactive and thus less stable and a range of other metals including aluminium, which gives enhancement with UV excitation, platinum, lithium and sodium have also been shown to work. Many different types of roughened surfaces have been prepared including nano-engineered layers and particles, colloidal particles and structures, electrodes and cold-deposited metal island films.

Since SERS was discovered experimentally, many theories were proposed in the early stages of development. To some extent most of them contained an element of truth. As we will see later in this chapter, the true nature of the theory is still an active research field but there is now a general consensus on the basic points backed by many years of accumulated experimental evidence.

The key part of the enhancement is due to the interaction of the analyte with collective oscillations of electrons on the surface of the metal called surface plasmons. On the surface, the metal atoms which control the properties of the loosely attached conduction electrons in the bulk of the metal are only on one side. As a result, there is electron density some distance from the surface and there is potential for facile movement of electrons in a lateral direction along it. When radiation interacts with these electrons, they can begin to oscillate as a collective group across the surface. These oscillations are termed surface plasmons. When created by visible or NIR excitation, they extend into the first few layers of the metal surface and the nature of the metal and the surface geometry control the frequency, frequency range and intensity of the plasmon formed. It so happens that both silver and gold plasmons oscillate at frequencies round about the visible region and therefore, they are suitable for use with the visible and NIR laser systems commonly used in Raman scattering. On a smooth surface the oscillation occurs along the plane of the surface and no light will be scattered. To obtain scattering, there needs to be an oscillation perpendicular to the surface plane and this is achieved by roughening the surface. This locates more electron density in the valleys of the roughened metal surface and scattering is caused as the oscillating electrons move up towards the peaks creating a component of the oscillation perpendicular to the surface. The nature of this roughness is also involved in controlling the frequency and frequency range at which the plasmon oscillates.

Metals can both scatter and absorb radiation but the ratio of the two is dependent on the metal and on the excitation frequency. With excitation in the middle of the visible region, the ratio for silver favours scattering more than that for gold, with the ratio for gold improving as excitation is moved towards the NIR. In the basic theory of the interaction of light with a material, the dielectric constant of the metal is divided into real and imaginary parts. Scattering is associated with the real part and absorption with the imaginary part. Some SERS papers use this terminology but it is largely beyond the scope of this book, however, further information can be found in Ref. [4]. Practically, one simple way of characterizing a plasmon is to measure the absorption spectrum. This is particularly straightforward if the substrate is a colloidal suspension since a spectrum can be obtained from a cuvette containing the suspension. However, as the beam passes through the sample it will both absorb and scatter. Most scattered light will not reach the detector and so will be measured as absorption. Thus, the spectrum obtained is not an absorption spectrum but an extinction spectrum consisting of absorption and scattering components. This is useful, but to obtain the separate scattering and absorbance components more has to be done. For example, using a fluorescence spectrometer the scattered light component can be obtained by collecting the light at 90° to the input beam. However, depending on the optics, only a fraction of the scattered light is collected. Usually the input beam is broad, illuminating the sample to some depth so that some scattered light can be self-absorbed by the colloid or re-scattered before it is detected. A correction may need to be made for this.

In addition to the ratio of absorption to scattering, the nature of the roughness is important. Usually for a metal surface made, for example, by electrochemical roughening of an electrode or by depositing the metal onto a surface, there are many different roughness features of varying dimensions each with a specific resonance frequency and a natural band width. The result of this is that the plasmon on the surface usually covers a broader range of wavelengths than predicted for a single roughness feature. In addition, as discussed later, areas of very intense scattering known as hot spots occur where specific features produce high fields and hence high intensities.

In summary, to obtain good SERS, it is best to use a material which will provide a suitably roughened surface which is time-stable for the period of the measurement. It needs to be of a material for which plasmons can be induced in a frequency range which includes that of the exciting laser and it is best if the ratio of scattering to absorption is favourable. In addition, there must not be too thick an oxide or other barrier layer between the adsorbed analyte and the surface. The basics of the theory behind the enhancement are given below.

5.2 ELECTROMAGNETIC AND CHARGE TRANSFER ENHANCEMENT

Historically, two different theories of surface enhancement emerged [4–8]. In one, the analyte is adsorbed onto, or held in close proximity, to the metal surface and interacts with the surface plasmon to give enhanced scattering. This is called electromagnetic enhancement. In the other, called charge transfer (CT) or chemical enhancement, the adsorbate chemically bonds to the surface. The original theory for CT enhancement proposed that the exciting radiation is absorbed by the metal causing hole pair formation. The enhancement is produced by transfer of energy from the hole pair to the molecule to cause the Raman process and by the return of the reduced energy back to the metal before scattering from it. Most theoretical studies now agree that electromagnetic enhancement is the dominant process but where an adsorbate chem-isorbs, the new surface species created can significantly increase the enhancement. This means that the bonding part of the CT theory is still relevant but enhancement by light absorbed by the metal seems to be weak. Therefore, SERS can be considered as a single process based on enhancement involving the plasmon, provided account is taken of any new bonds formed at the surface.

In the analysis of the original pyridine data, the electromagnetic enhancement was estimated at about 10^4 and the CT enhancement at 10^2. This will vary depend-ing on the molecule. Since CT enhancement requires that the adsorbate is bound to the metal, it only occurs from the first layer of the analyte attached to the surface whereas electromagnetic enhancement will also occur from second and subsequent layers. Early experiments carried out under clean conditions using spacers to separate layers clearly indicated a very large enhancement of scattering, more than would be expected from electromagnetic enhancement, from the first layer in direct contact with the metal [9]. Figure 5.2 shows this for the symmetric stretch in benzene. Apart from the first layer, cyclohexane layers are added to space the benzene from the surface.

5.2.1 Electromagnetic Enhancement

The simplest description of electromagnetic SERS is for a molecule adsorbed on a small metallic sphere. When the sphere is subjected to an applied electric field from the laser, the field at the surface is described by

$$E_r = E_0 \cos\theta + g\left(\frac{a^3}{r^3}\right)E_0 \cos\theta \qquad (5.1)$$

E_r is the total electric field at a distance r from the sphere surface,
a is the radius of the sphere,
θ is the angle relative to the direction of the electric field,
g is a constant such that,

$$g = \left(\frac{\varepsilon_1(v_L)-\varepsilon_0}{\varepsilon_1(v_L)+2\varepsilon_0}\right) \qquad (5.2)$$

ε_0 and ε_1 are the dielectric constants of the medium and of the metal sphere, respec-tively. v_L is the frequency of the incident radiation.

At some point where the denominator is at a minimum, the value of g will be at a maximum. ε_0 is usually close to 1 and consequently this usually occurs when ε_1 is equal to -2. At this frequency, the plasmon resonance frequency, the excitation of

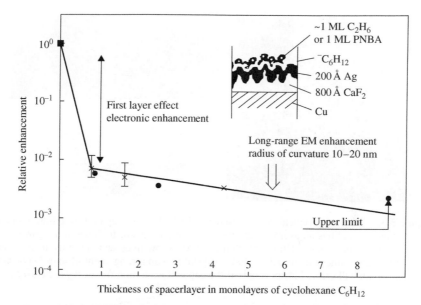

Thickness of spacerlayer in monolayers of cyclohexane C_6H_{12}

Figure 5.2. An early experiment carried out under clean conditions in which benzene was adsorbed onto a roughened silver surface to give the first layer and cyclohexane was added to provide spacing layers between the surface and the benzene in subsequent layers. The intensity plotted is the symmetric stretch of benzene. It clearly shows the importance of the first layer. Source: Taken with permission from Mrozek, I. and Otto, A. (1989). *Appl. Phys. A: Mater. Sci. Process* **49**: 389 [9].

the surface plasmon greatly increases the local field experienced by the molecule absorbed on the metal surface. In essence, the adsorbed molecule is bathed in a very free-moving and oscillating electron cloud, the plasmon, and this induces polarization in the molecule. At the metal surface the total electric field is averaged over the surface of the small sphere. At any point on the surface the electric field may be described by two components, the average field perpendicular to the surface and the average field parallel to the surface. It is the magnitude of the field perpendicular to the surface which is important for SERS.

Few real particles are truly spheres and the greatest single particle enhancement generally occurs at projections or on the points of shaped particles such as rods and stars. Figure 5.3 shows a transmission electron micrograph (TEM) of particles from a silver colloid which has been dried out on a surface. Unless great care is taken, there are usually some size as well as shape differences between particles as the high-resolution images show and these will have plasmons at different frequencies. Most measurements are taken in a volume or area which includes many particles and so the overall plasmon will be broadened and may have some separate features showing in it. Plasmon enhancement will then occur over a longer frequency range but with different particles and structures contributing at each excitation n frequency

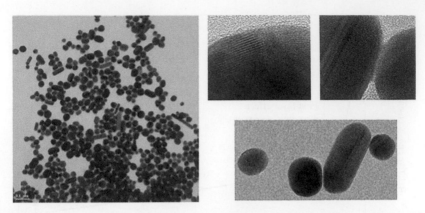

Figure 5.3. TEM of a typical colloid used in SERS. It was prepared by citrate reduction of silver nitrate and has an average particle size of about 35 nm. It shows some particle size variation and some needles. Higher resolution TEMs show lines which are from silver atom layers and thicker lines due to defects in the crystal and an arrangement of gaps between the particles which will create hot spots only if the gap and excitation frequency are correct.

used. Further, where the experiment is set up to obtain enhancement from single or small groups of nanoparticles, structural variations will cause significant variation in enhancement with any one excitation frequency. However, as explained in the next paragraph, in many systems most of the enhancement arises from the interaction between particles so where this can occur, the arrangement as well as the individual particles needs to be considered.

In many systems the most intense scattering arises not from single particles but from regions where intense fields are generated in the gaps between closely spaced particles. These are called hot spots and they usually provide most of the enhancement [4, 5]. Thus, if the dried-out film shown in Figure 5.3 was used as a substrate, most of the enhancement will arise from regions within the film which give effective hot spots. A random array, like this, produces hot spots which have plasmons with a range of resonance frequencies but there are very good modelling studies particularly for simple cases such as dimers and trimers [10, 11]. The field strength is related to the size of the gap between the particles. Too narrow a gap and quantum tunnelling or conduction will even out the field and too wide a gap and the strength of the field is diminished. Broadly, a gap of about 1 nm gives the largest enhancement. The hot spots contain a number of particles and the resonance frequency for a hot spot will normally be at a lower energy than that for the single particles creating it.

Either with excitation over a larger area of an irregular film or with a large-enough interrogation volume in a colloidal suspension, many active sites scatter many times during the accumulation time of a second or more, creating an averaging effect that permits quantitation. However, at the single particle or single hot spot level, enhancement can be very variable.

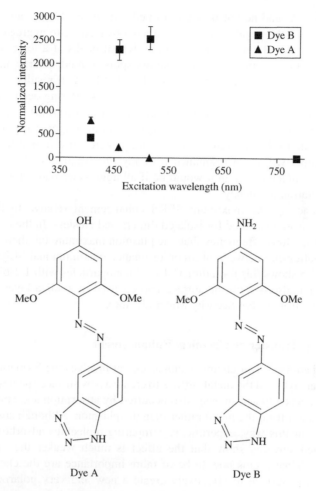

Figure 5.4. Plot of the intensity dependence of one peak for both a non-aggregating dye (A) which will give single-particle enhancement at about the plasmon maximum of unaggregated colloid (406 nm) and an aggregating dye (B) which will give enhancement mainly from clusters containing hot spots with lower energy resonance maxima. Source: Reprinted with permission from Faulds, K., Littleford, R., Graham, D., et al. (2004). *Anal. Chem.* **76**: 592 [12]. Copyright 2004 American Chemical Society.

Figure 5.4 shows the intensity of the SERS from a major peak for two dyes designed specifically for SERS. They bind to the surface through the benzotriazole ring system shown at the foot of the dye structures in the figure. They differ in that one has a hydroxyl group present as O^- at neutral pH so that on adsorption to the silver particles the negative charge on the particle surface will be maintained. Since it is the negative charge which keeps the particles singly suspended, no aggregation will occur (dye A) and no hotspots are formed. The other dye has an amine group which will reduce the negative charge as it adsorbs leading to aggregation and hot spot formation (dye B). A broader plasmon, which peaks at a lower energy, will form with dye B due to the many different hot spots formed. Charge and size measurements confirm this. Figure 5.4 shows the SERS which results with different excitation wavelengths. With dye A the most intense scattering occurs at about the plasmon resonance frequency of unaggregated colloid (406 nm) and with dye B greater enhancement occurs at lower energies (i.e. at longer wavelengths). Bear in mind that due to the different resonance frequencies of different hot spots only a limited number will be in resonance with any one frequency of excitation. The effect shown is much less than it would be if all aggregates had active hot spots at the same excitation frequency.

The frequency of the maximum SERS enhancement relative to the plasmon maximum has been measured for isolated dimers and clusters. In the clearest cases, it is obtained at a lower frequency than the plasmon maximum but there seems to be no simple relationship between plasmon resonance maximum and SERS maximum [13]. Figure 5.5 shows this for a dimer of gold nanoparticles with 1,2-bis(4-pyridyl) ethylene (BPE) adsorbed on the surface and encased in a silica layer for stability ('nanoantenna' from Cabot Security Materials, Inc).

5.2.2 Charge Transfer or Chemical Enhancement

As discussed above, CT or chemical enhancement involves the formation of a bond between the analyte and the metal surface to create a new surface species. Originally, the theory proposed that scattering also occurred by absorption and emission of the exciting radiation from the metal rather than the plasmon. Although some enhancement may occur this way, experiments comparing molecules adsorbed on smooth and roughened surfaces show that the effect is much weaker than the plasmon enhancement. What seems now to be of more importance are the chemical bonds formed with the surface to effectively create a new and very polarizable surface species. Whether such a bond forms and how strong it is, is very much dependent on the analyte and the specific surface so CT may be expected to vary significantly between analytes.

Figure 5.6 illustrates the basic concept for one possible arrangement. The molecule on the surface will have a highest occupied molecular orbital (HOMO) and a lowest unoccupied molecular orbital (LUMO) at specific energy levels. On the surface, the metal will have bands spanning a range of energies with the HOMO half filled with electrons. By changing the potential, the energy of the highest energy electrons at the surface can be altered to match one of the levels on the adsorbate making energy transfer much easier and altering the electron distribution across the surface, thus affecting the polarizability. Some analytes are Lewis bases and may donate electrons so that CT transitions can be either from the metal to analyte or from analyte to metal depending on the energy levels and electrode potential [14].

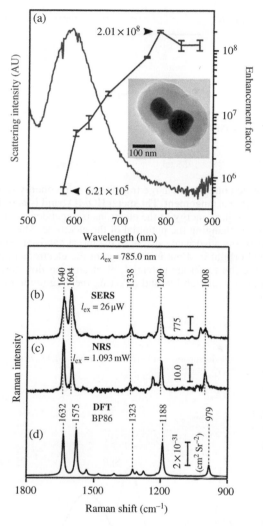

Figure 5.5. (a) The intensity of the 1200 cm^{-1} band of BPE plotted against wavelength for a dimer of gold nanoparticles coated with BPE and then a silica shell (blue) and the plasmon resonance for this dimer (red). (b)–(d) The **SERS** spectrum, the Raman spectrum and the calculated spectrum. This clearly shows that the maximum enhancement is at a lower energy than the plasmon maximum. Source: Reprinted with permission from Kleinman, S.L., Sharma, B., Blaber, M.G., et al. (2013). *J. Am. Chem. Soc.* **135**: 301 [13]. Copyright 2013 American Chemical Society.

This approach has been used for many years and since it is the approach taken in much of the literature it is repeated here. However, modern calculations can model the actual bonding involved at the surface so that the effect of potential can be calculated using new molecular orbitals formed when the adsorbate and surface metal ions mix, as described below and illustrated in Figure 5.7. The effect of potential on

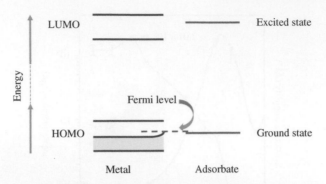

Figure 5.6. The basic concept of charge transfer showing energy levels for a metal and adsorbate in one possible arrangement. The metal HOMO band is half filled with electrons but at the surface the highest energy of the electrons is dependent on the surface potential which can be varied by changing the potential of the electrode. The energy of the HOMO and LUMO orbitals of the adsorbate relative to the metal bands are set arbitrarily here so that the highest filled orbital is about in the range of the electrons in the HOMO band of the metal. By varying the electrode potential, electrons can flow towards or away from the adsorbate, changing polarizability and affecting the magnitude of the charge transfer contribution to SERS.

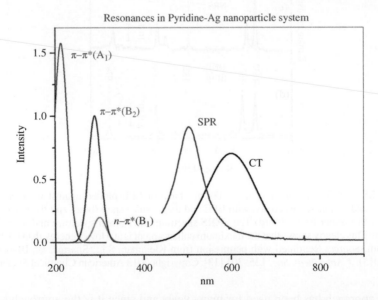

Figure 5.7. Resonances for pyridine including the charge transfer term (CT) as well as the surface plasmon resonance term (SPR) and molecular states which could contribute depending on excitation frequency. Source: Reprinted with permission from Lombardi, J.R. and Birke, R.L. (2009). *Acc. Chem. Res.* **42**: 734 [14]. Copyright (2009) American Chemical Society.

the polarizability of these new, very polarizable states can then be calculated. When possible, this is a better way to describe CT.

Lombardi and Burke added an extra term or resonance to the electromagnetic theory to account for CT [14]. The theory essentially uses similar mathematics to that used to define B-term enhancement in Chapter 4 by creating a CT state by mixing in orbitals of higher energy. Thus, in addition to the electromagnetic term generated by the surface plasmon and any molecular resonance contributions from the molecule, a new term is created to account for CT by mixing the excited states including those from the HOMO and LUMO of the metal. Figure 5.6 shows this for pyridine. The π–π^* transitions in the near-UV for pyridine, which could contribute molecular resonance and the plasmon resonance (in SPR – S stands for surface), are shown along with a third term to account for CT. In Figure 5.6, when the laser excitation is chosen to be towards the red, two terms (SPR and CT) should dominate but in some cases, say when a dye is the analyte, there will be π–π^* transitions in the visible region and molecular resonance will also contribute to resonance.

This provides a general approach but CT or chemical enhancement implies specific bonding, the nature of which will depend on the substrate and the adsorbate used. The bond formed will be species specific and the properties of the surface layer such as pH, ionic strength and dielectric constant will be important as will any packing forces between closely spaced adsorbates. With some simplifications it is possible to calculate the nature of the surface species formed and so show the influence of the metal on the molecular states directly. Figure 5.8 shows the result of a calculation of the bonding which results when pyridine is added to a gold cluster. Pyridine adsorbed on a metal surface will form new bonds with the metal atoms directly bonded to the adsorbate and cause a perturbation of the electronic structure of the surrounding metal atoms with the magnitude of the effect reducing with distance from the bonded atom. This makes an adsorbate metal atom cluster a reasonable model to calculate a CT state. The extended orbitals into the metal surface make the dipole across the new species very sensitive to the electrode potential and the effect can be calculated. The results in Figure 5.8 show that modern calculations can make reasonable predictions of SERS and the whole area is better understood [14, 16] but they are a long way from perfect. For example, SERS is usually carried out under water in a surface environment with a specific ionic strength and pH and any calculation must take this into account. In addition, bonding may distort the surface metal layer, the adsorbate may bond at an angle and packing forces between molecules may have an effect.

Thus, the basics of SERS theory are now fairly clear and the technique is used with some confidence. However, it is likely that there will be more development to come. We now know that hot spots can be generated from very few atoms as pointed out when describing Tip-enhanced Raman scattering (TERS) in Chapter 7 and in other systems local plasmons generated in small volumes have been postulated to give SERS. In addition, as the section on surface chemistry below will demonstrate, the nature of the bonding between an analyte and a surface can be more complex than simply the addition at a specific angle of a monodentate ligand like pyridine.

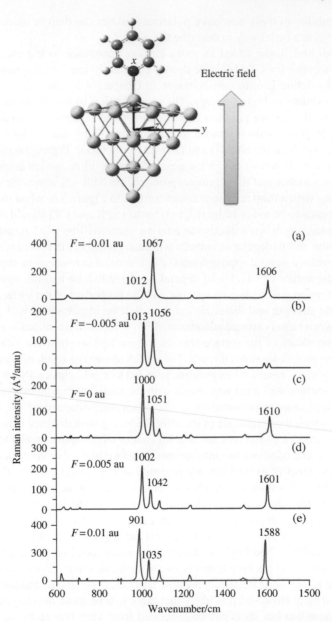

Figure 5.8. Diagram of pyridine adsorbed on a gold 20 cluster together with the spectrum calculated at different surface potentials. Source: Reproduced with permission from Zhao, X. and Chen, M. (2014). *J. Raman Spec.* **45**: 62 [15].

5.2.3 Stages in the SERS Process

SERS results from events on two different distance scales. In Figure 5.9 one wavelength of the exciting radiation at, say, 500 nm is represented as a red horizontal line which also indicates the direction of propagation. Perpendicular to this, the oscillating dipole of the radiation, which is too large to show, induces the plasmon, the direction and approximate wavelength of which is shown by the blue line. This scale is much bigger than that of any molecular process and processes on this scale are called far field. The second part of the process occurs on a molecular scale and close to the adsorbed molecule. A local field at the point where the molecule is adsorbed interacts with the molecule (red in the expanded Figure 5.9b) inducing the polarization change. Events on this scale are called near field. Thus, the scattering process occurs on two scales, the far-field part involving the formation of the plasmon and the scattering after the Raman event, and the near-field part involving the induction of the polarization on the molecule by the near field, the Raman event and the return of the reduced energy to the plasmon (green arrows in Figure 5.9b). Thus, the scattered radiation will be at a lower frequency than the

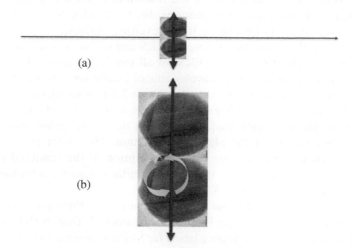

(a)

(b)

Figure 5.9. An illustration of the different dimensions involved in the production of SERS. In (a), the red line indicates the length of one wavelength of the exciting radiation relative to the nanoparticle and the direction of propagation but not the amplitude of the wave. The blue line indicates the direction and approximate magnitude of the plasmon induced at 90° to the direction of the light. The magnitude shown is roughly the distance from the surface where there is still any significant electron density. This dimension is called far field. In (b) the pair of particles is shown on an expanded scale, with the position of an adsorbed molecule (red) and the transfer of energy to and from it (green arrows) to create the enhanced Raman scattering. This dimension is called near field.

plasmon resonance frequency. The energy transfer to and from the molecule are both proportional to the square of the frequency of the exciting radiation giving a fourth power dependence for the entire process. In passing, this diagram uses a high-resolution TEM of a single silver particle from citrate colloid which has simply been copied to create the dimer. However, it illustrates that the particle is structured and not a sphere.

5.3   SURFACE ENHANCED RESONANCE RAMAN SCATTERING (SERRS)

Very high enhancements were reported early in SERS development and the first reports of single-molecule SERS were made using dyes and, in particular, rhodamine where the enhancement factor was reported as about 10^{14} [17, 18]. Others have estimated both a lower number at about 10^{12} and a higher number of about 10^{15} or more but either way this is a huge enhancement and is best explained as due to a combination of the resonances shown in Figure 5.7 with CT and molecular resonance contributions due to an allowed transition of the molecule being of a similar energy to the exciting radiation. Some papers do not acknowledge the distinction between SERS and surface enhanced resonance Raman scattering (SERRS) using SERS for all molecules under all conditions. However, studies using different wavelengths of excitation indicate that the dye can have a specific effect. As an example, Figure 5.10 shows that at low concentrations when no aggregation is expected, a dye with an absorbance maximum different from the excitation frequency gave a maximum enhancement at the absorbance maximum not the plasmon maximum. The effect shown is quite dramatic and may in part be due to self-adsorption of the scattered radiation by the plasmon but it does show that the molecular resonance term has a large effect which will not occur with SERS.

One key advantage of SERS and SERRS is that when a fluorophore is adsorbed on a metal surface the fluorescence is effectively quenched. Thus, both fluorophores and non-fluorophores can be effective labels for SERRS giving a much larger selection of dyes than for resonance Raman scattering or fluorescence. In addition, due to the high enhancement factors, very low concentrations are often used so that self-absorption of the scattering by dye in solution is less of a problem than with resonance Raman scattering.

The use of two terms, SERS and SERRS, can make reporting unnecessarily complicated since in many studies at one wavelength the SERS/SERRS active label is just a tag and the difference between the two is of little consequence. In other studies, it is important, such as in investigations involving the fundamentals or wavelength dependent on SERS. In this book, the term SERS will be used except in cases where the difference is important.

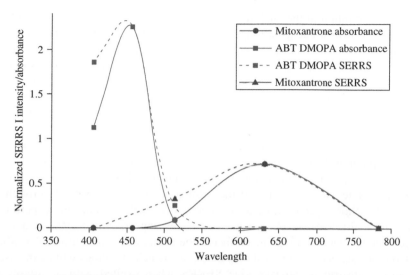

Figure 5.10. SERS/SERRS at four different wavelengths for two dyes adsorbed at low concentrations on suspended silver particles. There was no evidence of any aggregation and the plasmon resonance was measured at 406 nm. ABT DMOPA (dye B in Figure 5.4) has an absorbance maximum at 453 nm and shows maximum SERS intensity near this frequency and to the level of accuracy possible here. In contrast, mitoxantrone gives a maximum intensity at about 600 nm close to where the dye absorbance maximum is and away from the plasmon resonance maximum. The position of the CT term is not known. Source: Reproduced from Cunningham, D., Littleford, R.E., Smith, W.E., et al. (2006). *Faraday Discuss.* **132**: 135 [19] with permission from The Royal Society of Chemistry.

5.4 SELECTION RULES

SERS spectra are not straightforward to interpret. New bands not seen in spectra from normal Raman scattering can appear in SERS and some bands, which are strong in normal Raman scattering, can become weak or disappear in SERS. There can also be changes in the spectrum as the concentration on the surface changes. Since scattering efficiency is dependent on the induced molecular polarizability component perpendicular to the surface, the angle of the adsorbate will critically affect intensities. For example, if an analyte such as pyridine or mercaptobenzoic acid is adsorbed so that the plane of the aromatic ring is perpendicular to the surface, the scattering process will be efficient since some of the strong in-plane C—C vibrations will have displacements perpendicular to the surface creating a polarization component vertical to it. If the plane is parallel to the surface, these modes will give a much smaller polarizability change perpendicular to it. However, some modes and the totally symmetric mode, in particular, will give some scattering because during the vibration the electron density will be squeezed as the molecule contracts causing a polarization change vertical to the surface.

The concentration of the analyte can affect intensity. If there is plenty of surface available, the lowest energy arrangement of the analyte and the metal will determine whether it lies flat, perpendicular or at another angle. However, if the surface is crowded, the analyte may reorientate to take up as little surface space as possible with packing forces between molecules contributing to the overall stability to make this the lowest energy situation. Thus, the concentration dependence of SERS at near monolayer coverage can be nonlinear and give significant changes in relative band intensities.

One reason new bands can appear in SERS is that with a centrosymmetric molecule adsorption of the molecule onto a metal surface will effectively break the centre of symmetry [7]. This results in the mutual exclusion rule (see Chapter 3) no longer being applicable, allowing some of the infrared-active bands to break through and appear in the SERS spectrum. Pyrazine, which was widely used in early experiments, shows infrared-active and Raman-inactive bands appearing in SERS. In addition, where chemisorption is a significant factor, the nature of the new molecular orbitals created in the surface species will help determine which vibrations will be intense.

Centrosymmetric porphyrins give SERRS when excited with a suitable frequency but the changes compared to the equivalent resonance Raman scattering are smaller than for pyrazine and this is a common feature of SERRS. The strong molecular resonance contribution reduces the effect of orientation on the spectrum. It can still be observed as was shown in a study of cytochrome c from below to above monolayer coverage. As explained in Chapter 4 for resonance, the symmetry results in intensities which are very dependent on excitation frequency. In the visible region, hemes have both a forbidden and an allowed electronic transition. Excitation in the energy region of the allowed Soret band produces strong scattering from totally symmetric A_{1g} vibrations and excitation in the region of the forbidden transition produces scattering from B_{1g} and B_{2g} modes and again intensities vary between molecular resonance and SERRS spectra. However, when the chromophore is present in a protein, the protein structure usually prevents direct contact with the surface so there is no CT contribution.

Early in the development of SERS, Creighton [7] proposed selection rules. They refer to physisorbed molecules and do not take into account any new species formed by chemisorption. In practice, to explain qualitatively why some bands are intense they are effective in most cases. The appearance of new bands and the disappearance of existing ones can make it difficult to relate the spectra obtained on the surface to that obtained from normal Raman scattering. In addition, since the sensitivity of SERS compared to normal Raman scattering is huge, the dominant features of a spectrum could arise from a trace contaminant which adsorbs strongly on the surface and has a high enhancement factor. In many cases, salient bands can be recognized and the analyte identified and discriminated from other species in a mixture. However, if an unusual band appears, care must be exercised in the assignment since it could arise either as a result of SERS selection rules or due to a small amount of contaminant.

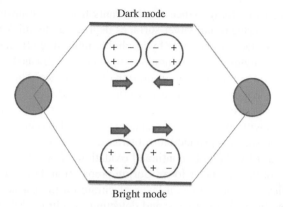

Figure 5.11. Representation of the coupling of the wave functions for the surface electrons from two particles to form two new plasmons, the lower energy bright mode and the higher energy dark mode.

As already discussed, when dimers or clusters are formed, the plasmon frequency is shifted. Figure 5.11 shows the wave functions for the surface electrons of two particles coupled to form two new plasmons, a bright mode which causes scattering and is at a lower energy than for the single particle and a higher energy dark mode which if the particles are identical will not scatter. However, scattering can be obtained from the dark mode if the particle size differs so that the symmetry is reduced. This interaction between particles needs to be considered if the wavelength dependence of a SERS hotspot is to be considered.

5.5 SURFACE CHEMISTRY

An understanding of the chemistry at the surface is useful to help set conditions to obtain effective adsorption of an analyte and to help predict the nature of any CT complexes formed. Often, the metal surface is under water and in the presence of dissolved oxygen and electrolytes such as sodium chloride. With gold, the surface of the gold will be coated with an adhering layer which contains solvent and electrolytes giving the surface distinct characteristics such as altered dielectric constant, pH and charge, all of which will affect analyte adsorption. Some analytes will adsorb on the surface in a nonspecific way due to the properties of the surface layer. Where chemisorption occurs, an analyte with a specific attachment group such as a thiol will bond to the gold atoms and some charged species will bond by association with the charged surface. Obviously, clean gold and silver substrates can be made and used in a vacuum when the above considerations do not apply, but these surfaces are often prepared in a vacuum and used later in air with solutions. In this case they should be vacuum packed for storage since contaminants from the atmosphere often give signals from these surfaces and silver surfaces degrade quite quickly. We have also found some gold surfaces to be quite hydrophobic inhibiting efficient adsorption of weakly attaching analytes. One common way to obtain spectra as discussed later is to increase concentration or dry out more solution on the surface. This usually produces uneven multilayers which give Raman scattering not SERS.

The surface of silver is much more reactive. Both under water and in air the surface will be coated with oxide unless great care is taken. This can be a problem in the production of time-stable reproducible substrates. Silver colloid made with sodium borohydride is short lived due to continued oxidation. However, stable colloids can be made when a protecting layer forms during preparation. For example, silver colloid made by citrate reduction has a layer of citrate on the surface as shown both by SERS and by displacing the layer and detecting the citrate. Silver (I) citrate is a polymer with some negatively charge groups present due to unattached carboxylate groups as shown in Figure 5.12 and this polymeric layer is quite stable. Other stable colloids such as those made with EDTA or hydroxylamine probably have similar layers but more is known about the citrate system. Three different general ways an analyte can bond to the citrate-coated surface are shown in the figure. The citrate layer can prevent many but not all negatively charged analytes from bonding directly to the surface. They can still bond more loosely by hydrogen bonding or polar bonding to positive ions in the surface layer. In this case a higher concentration of analyte in solution will help force surface adhesion at the expense of some sensitivity. If amines are protonated and therefore positively charged, as is often the case at neutral pH, they can bond to the negative surface through a polar bond. Alternatively, unprotonated they can bond directly to silver (I). Thiol groups form a strong bond to

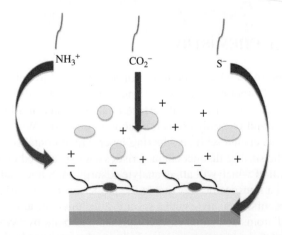

Figure 5.12.   Three general ways an analyte can interact with a silver citrate colloidal surface. The green band represents silver with the grey band representing silver oxide. The layer above is citrate with the silver (I) ions in the polymeric layer represented by the blue ellipses.

silver (I). They may be negatively charged if ionized as shown in the figure but the strong bonding will strip off the citrate.

If the analyte does not adhere strongly, the surface can be changed, for example, by adding a ligand which bonds strongly to the surface and changes the surface charge from negative to positive or by forming a layer such as a lipid layer with a different dielectric constant on the surface. It is also possible to use a clathrate which will take the analyte into the structure and itself adsorb on the surface.

5.6 SUBSTRATES

There is a plethora of effective substrates and more are constantly being reported. Essentially any roughened surface of a suitable metal will give enhancement but much more care is required to get high, reproducible and stable enhancement. A few general types include colloidal suspensions, designed surfaces, stable nanotags, rods, stars and other shaped particles, hollow gold nanospheres and rough metal films. Each of these has its own unique advantages. The silver particles shown in Figure 5.3 are typical of the types of particles found in many suspensions although suspensions with fewer variations can be prepared with care. Both silver and gold colloids can be made with different sizes and therefore different plasmons. Single-particle SERS has been obtained but, as described above, much larger enhancements are usually obtained if the particles are aggregated in a controlled fashion to obtain hot spots. Colloid can be time stable and last for years if properly made. What is required is a relatively high surface charge which can be measured by such techniques as zeta potential. Aggregation to create hot spots is most commonly achieved by reducing the surface charge using inorganic salts such as sodium chloride or magnesium sulphate or organic agents such as positively charged poly-L-lysine or spermine. Aggregated suspensions are not stable, with the aggregates growing with time and eventually precipitating out. Thus, the degree of aggregation induced should be controlled so that the suspension is stable for longer than the measurement period as in the examples of quantitation described later in this chapter. The advantages of using colloid include high enhancement factors, direct use of solution methodology such as titration, a fresh surface for every measurement and easy removal of excess analyte, for example, by centrifugation and resuspension. In addition, only an analyte attached to the surface will give a strong signal whereas with a substrate, the deposited analyte can create multilayers which may give Raman scattering rather than SERS.

Solid surfaces have the advantage that they are very compatible with use in a microscope. Simply add a drop of liquid to a preprepared substrate and focus on the surface. Designed surfaces are aimed at providing reproducible and reliable enhancement both across the surface and between each slide. They can be created by using theory to calculate the roughness required to obtain a plasmon of a specific frequency and manufacturing a device with controlled sizes [20]. As an example,

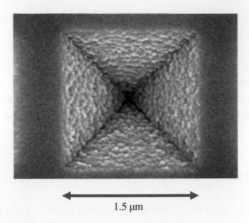

1.5 μm

Figure 5.13.   One of the pyramidal pits in Klarite which form a regular surface array with a specific local plasmon at each pit determined by the dimensions. The rough gold layer which coats the pit and provides a more local field in each pit can be seen.

Klarite, which, at the time of going to press, is no longer available commercially, achieved this by using semiconductor technology to create a silicon surface with a regular array of pyramidal shaped pits when coated with gold which gave a regular local plasmon at each pit. The interior of each pit was coated in a rough layer of gold to give intense local fields within the pit (Figure 5.13). This degree of control gave good reproducibility across a surface and also between devices. Substrates can now be produced using electron lithography to achieve the necessary control of the surface morphology. Films can be made to produce high enhancements, but for widespread use, reproducibility between devices and across the surface is important. Further, the plasmon may be broadened to give lower enhancement but work over a wider range of excitation frequencies. Klarite was tuned to be effective over a range of wavelengths, to be reproducible and to be time stable.

Many surfaces are made simply by techniques such as etching or cold deposition of an incomplete film to form adjacent metal 'islands' or by immobilizing colloidal particles on a surface. The enhancement across this type of surface and between different samples can vary quite widely depending on how well the surface can be made. Solid substrates have the advantage that they can be prepared and used when required. However, they are difficult to clean, so they are usually 'one off' devices, can age and can be expensive if techniques such as electron lithography are used. One type of surface that can be cleaned effectively are electrodes which can be cycled to provide a new surface for every measurement.

Care must be taken using solid substrates. Most SERS is from analyte present in the first layer on the surface but if a drop of solution is added to a surface and dried out, there is often a 'coffee ring' effect with a relatively dense rim of material in a ring and a much less coverage in the middle. This is not helped by the fact that many bare metal films tend to be hydrophobic. A common practice is to focus on the thicker layers in the coffee ring. Further, even if the analyte distribution is even, poor signal-to-noise spectra may tempt the analyst to add more material. Spectra obtained from thick layers are usually due to normal Raman scattering from multilayers present on the surface rather than SERS which will be mainly from the first layer and easily lost by interference between layers and self-absorption of the scattered light. Dipping the film into a solution, and then washing it is a better approach in that the analyte must have been positively attracted to and must adhere to the surface to be detected and more uniform coverage is obtained. However, even with this approach, a number of washings should be performed and the spectra recorded after each since adhering non-adsorbed material in solution on the wet surface can still accumulate as the film is dried. To test for SERS, where possible, equivalent amounts of analyte can be applied to rough and smooth metal films and the signals compared. Mercaptobenzoic acid is commonly used as a test analyte. It is water soluble and adheres strongly to gold and silver surfaces through the thiol group, so even rigorous washing leaves a surface layer.

A thin metal layer can be directly deposited on the analyte, for example, by cold deposition. If the layer is sufficiently thin, the surface will be rough and the plasmon will oscillate through the deposited layer and be active on both sides. Thus, excitation on the side away from the analyte can cause the plasmon to interact with the surface of the analyte. Collection on the side away from the analyte can give SERS from the surface. The metal layer will prevent most or all of the exciting radiation being directly transmitted through the film and will certainly prevent the weaker Raman scattering being transmitted back through it providing very good discrimination between SERS and normal Raman scattering. For example, good SERS was obtained from the polymer polyethylene terephthalate (PET) when an approximately 15 nm film of silver was deposited on the surface, the sample excited and the scattering collected from the silver side away from the polymer. The spectra (Figure 5.14) are different from the Raman spectra, due to SERS selection rules. In particular, new bands appear in SERS and the band above $1700 \, cm^{-1}$, which is due to the carbonyl group is weaker in the SERS spectrum than the Raman spectrum. In addition to being a useful analytical technique, this example illustrates two key features of SERS which were discussed in the theory section. First, scattering on the side away from the substrate would be expected with a very thin film with plasmon enhancement but difficult to explain with any other theory and second, there are different selection rules for SERS.

Some analysis methods use tags such as metal particles with a SERS/SERRS-active label adsorbed and coated with a protective layer to reduce reactivity and increase stability. The most common coating is to make a silica shell round a nanoparticle coated with a SERS label [22] as was used to obtain the spectra shown in Figure 5.5. The coating stabilizes the particle giving it a long lifetime and inhibiting chemical attack on the surface. This type of tag can be obtained commercially or made in-house but a successful outcome depends on being able to obtain complete surface coverage with the silica and retain single discrete tags. The tags can then be functionalized

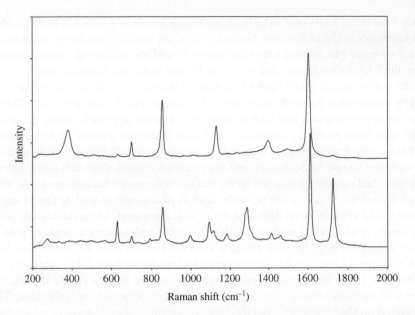

Figure 5.14. SERS spectrum of polyethylene terephthalate (PET) taken from a sample with a silver film cold-deposited on the surface (top) compared to a Raman spectrum of the same sample (bottom). SERS was recorded using excitation and collection from the side of the silver film away from the polymer layer. Source: Reproduced with permission from McAnally, G.D., Everall, N.J., Chalmers, J.M., and Smith, W.E. (2003). *Appl. Spectrosc.* **57**: 44 [21].

so that a biological molecule can be attached. This enables the tag to be used as a label in reactions such as those in antibody assays. There are other ways to achieve similar results. Small clusters of labelled silver particles can be made into beads with the cluster in the centre using organic polymers [23] or into particles (or tags) functionalized using a polymer such poly ethylene glycol (PEG) and a label. The functionalized particle can then be attached to an appropriate molecule such as DNA or antibodies.

There are many other ways of making successful tags. Hollow metal nanospheres (HGN's) have been quite widely used [24, 25]. These can be created by coating a cobalt particle with gold then dissolving out the cobalt to leave a sphere. The plasmon created is dependent on wall thickness and size. The shape of a nanoparticle has a significant effect on the efficiency of scattering. Rod-shaped particles have two plasmons, one oscillating along the rod and one oscillating across it. The plasmon oscillating along the rod scatters very efficiently from the ends and for gold the size and shape of rods can be controlled to tune the plasmon. A 'universal' probe with three dyes resonant at different frequencies which gives good SERS with a range of excitation frequencies has been made [26]. Carrying this on, many different shapes give good SERS such as star-shaped particles which are very effective and

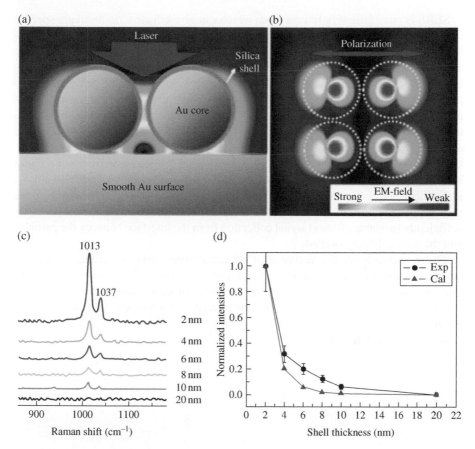

Figure 5.15. (a) Diagram of two Shiners on a smooth gold surface coated with pyridine and (b) a calculation of the induced polarization shown from above. (c) The spectrum from pyridine as a function of shell thickness and (d) a plot of intensity as the silica film increases in thickness. Source: Reprinted with permission from Li, J.-F., Zhang, Y.-J., Ding, S.-Y., Panneerselvam, R., and Tian, Z.-Q. (2017). *Chem. Rev.* **117**: 5002 [28]. Copyright (2017) American Chemical Society.

Ref. [24] reviews some of these. For those interested, it also goes much deeper into the plasmonics behind the enhancement including discussion of Fano resonances due to interference between broad and localized plasmon modes.

Finally, particles showing promise are SHINERS [27]. These are gold particles completely coated with a very fine layer of silica so that SERS-active analytes adsorbed on the outside can still be enhanced by electromagnetic enhancement from the protected particle. Figure 5.15 shows SHINERS added to a smooth gold metal surface to detect a molecule on the surface by creating a plasmon between the surface and the SHINER.

SERS is one of the very few methods which can give effective, molecularly specific information about a monolayer of an adsorbate on a metal surface, *in situ*, in aqueous solutions. However, this can be applied without the need for special particles. For example, benzotriazole is widely used as an anticorrosion agent for copper and an anti-tarnish agent for silver. Adding silver colloid to a copper surface treated with benzo-triazole gives good SERS and some information about the nature of the surface ligand interaction including, if the surface is not dried out, some information on the surface under water [29]. One difficulty in all these cases is knowing whether or not the analyte detaches from the analyte surface and adheres to the added particles, so the spectra obtained require careful interpretation. This technique has been extended to cover printing inks on paper, dyes in fibre, inks, pigments in artworks and other surfaces [30–32]. To make this work, the concentration of particles is critical. Too thin a layer and there are no hotspots, too thick a layer and the exciting and scattered radiation do not penetrate sufficiently to enable efficient signal collection from the interface between the particles and the material being studied.

The molecularly specific nature of SERS can also be used to make labels for product identification at a distance. For example, SERRS from a dye adsorbed on colloid and fixed in a polymer was detected at distances of up to 20 m within 10 seconds using a handheld Raman spectrometer connected to a telescopic lens using a laser providing 3.6 mW of 532 nm excitation (Figure 5.16). This is a good illustration of how sensitive SERRS can be.

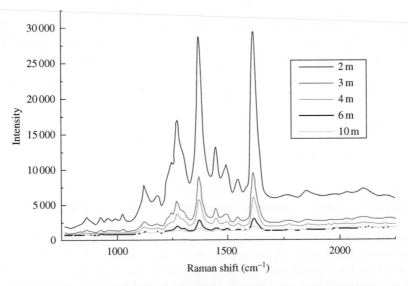

Figure 5.16. Distance dependence of the SERRS spectrum of a dye adsorbed onto silver colloid incorporated into a polymer. Source: Reproduced with permission from McCabe, A., Smith, W.E., Thomson, G., et al. (2002). *Appl. Spectrosc.* **56**: 820 [33].

5.7 QUANTITATION AND MULTIPLEX DETECTION

Where an analyte has a high SERS cross-section and adsorbs strongly to the substrate, simple, sensitive methods of analysis can be devised. Strong adsorption ensures that the analyte is concentrated on the surface aiding detection limits. The concentration of the coloured anticancer drug mitoxantrone is difficult to analyse in blood. It usually involves separation of the mitoxantrone from the blood serum and then chromatography. Using a flow system to control aggregation, a drop of plasma from a patient who had been undergoing a course of therapy with mitoxantrone was added to the flow cell. With visible excitation to give resonance or pre-resonance from the mitoxantrone, the plasma would be expected to give a fluorescence background. However, when the flowing stream containing the plasma was mixed with streams containing colloid and an aggregating agent thus diluting the plasma, excellent mitoxantrone spectra were obtained with little evidence of background fluorescence [34]. Figure 5.17 shows that there is good linearity over orders of magnitude of concentration. This encompasses the concentration range found in patients so that the technique could be used directly with no separation steps. This is an example where SERRS could be considered as the technique of choice. However, this assay used a coloured drug which adsorbed strongly onto the silver surface. Doing this with other drugs could be more difficult.

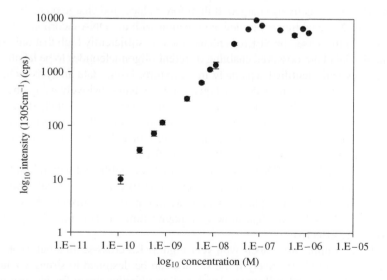

Figure 5.17. Intensity dependence of one peak from the SERRS spectrum of mitoxantrone in plasma. Source: Reproduced with permission from McLaughlin, C., MacMillan, D., McCardle, C., and Smith, W.E. (2002). *Anal. Chem.* **74**: 3160 [34]. Copyright (2002) American Chemical Society.

This example shows the effect of averaging discussed earlier. The focal volume was large enough for a number of clusters to be present in it at any one time and the accumulation time was long enough for sufficient scattering events to occur from each cluster to create an average result. In contrast, if the colloid is more heavily aggregated so that there are fewer but larger aggregates, the beam is very tightly focused and the accumulation time is reduced to 10^{-1} of a second, flashes of Raman scattered light can be observed as single large aggregates pass through the interrogation volume. There is less averaging in this situation and it is to be avoided if quantitative results are required.

Quantitation can be combined with multiplexing. DNA can be detected directly by SERS. The phosphate groups on the backbone of DNA present a negative surface so that adherence can be achieved by using positively charged surfaces. With a negative surface, single-strand DNA can adhere to the surface through the bases or through interaction with positively charged groups such as magnesium ions used to aggregate colloid. All four constituent bases can be discriminated and the relative concentration of each obtained [35, 36]. Dye-labelled oligonucleotides give SERRS a sensitivity that rivals fluorescence. This enables the use of DNA analysis procedures already developed for fluorescence but with the advantage of easier detection of a number of labels with no separation and less problem with variability (fluorescence can suffer from quenching or intensity changes due to relatively slight changes in experimental conditions). The number of detectable labels can be increased with the use of more than one excitation frequency. In the example described below, silver colloid was used as the substrate. It was aggregated with spermine, a polyamine which is protonated at neutral pH and so positively charged. It therefore reduces the charge on both the silver surface and DNA enabling controlled aggregation with the DNA incorporated into the aggregates. In this case, the SERRS enhancement is sufficiently high that only signals from the dye label are observed enabling different oligonucleotides to be labelled with different dyes and identified separately in a mixture. Using data analysis techniques, six different probes have been detected quantitatively and sensitively using one excitation wavelength [37]. The spectra obtained are quite complex and usually data analysis methods are used to quantitate each oligonucleotide using all the information in the spectrum but an example with three chromophores which directly relate to detection of disease shows the potential visually [38] (Figure 5.18).

A number of effective DNA assays have now been published which highlight the need to consider the surface chemistry. These systems can take clinical samples and after manipulation detect sequences which are diagnostic for disease. Many use silver colloid as the substrate. Spermine and magnesium chloride have been shown to be particularly effective.

The above example uses labels with very high SERS cross-section but weaker scatters can be measured effectively if the surface can be designed to strongly attach the analyte. Van Duyne and colleagues developed an effective assay for glucose using a surface made by depositing a rough metal coat on a regular array of polystyrene spheres of uniform size (FON's). To attach the glucose, initially a surface layer consisting of a mixture of decanethiol and mercaptohexanol was added to the silica. This gave good

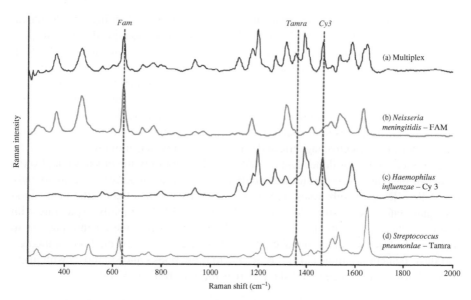

Figure 5.18. SERS of three oligonucleotides labelled with different dyes, FAM, Cy3 and Tamra. The oligonucleotide sequences are complimentary to sequences in viruses and can be used to detect them. The method used to do this was exoSERS (see Chapter 7) but here the reason for the figure is to illustrate that the SERS from the three separate sequences b, c and d can easily be distinguished in a mixture. Source: Taken from Gracie, K., Correa, E., Mabbott, S., et al. (2014). *Chem. Sci.* **5**: 1030 [38]. Published by the Royal Society of Chemistry.

results but was not as selective for glucose over fructose as was required and better selectivity has now been obtained using boronic acid [39, 40].

There are many other ways of developing successful detection methods. For example, cage-type molecules (clathrates) can be used to obtain SERS from molecules trapped in them [41]. However, effective SERS can be obtained even with weak adsorption. The detection limit will be affected but with a strong scatterer it can still be reasonable. Amphetamine can be detected from solution simply by adding it to aggregated silver colloid. However, centrifugation of the colloid and analysis of the supernatant by SERS showed that almost all of the amphetamine was still present in solution [42]. Further, flow systems and lab-on-a-chip systems can use SERS very effectively as a detection technique as further discussed in Chapter 6.

5.8 SUMMARY

SERS is a technique with huge potential giving exquisite sensitivity and the ability to positively identify an analyte in a mixture without separation. As a result, the field has grown rapidly in recent years assisted by the ready availability of handheld spectrometers which enable use in the field. The greater understanding, both technical and theoretical, has led to new methods which require more detailed description. As a result, techniques such as TERS, surface enhanced hyper Raman scattering (SEHRS), microfluidics, surface enhanced spatially offset Raman scattering (SESORS) and 3D cell mapping of particles are left to a later chapter. The examples given above were chosen to illustrate different ways of using SERS but a huge amount has been left out. For example, there are effective antibody assays available now.

Finally, the field was held back by the controversy over the existence and relative importance of the electromagnetic and CT terms, but now it is moving on. Ultimately, a true understanding is dependent on understanding not only the physics but also the nature of the bonding at the surface and this is still an area of active research not only for SERS. However, substrates are very much improved, the effect better understood and quantitation is possible. As a result, the use of SERS is rapidly expanding with many practical applications. This chapter is written to provide an introduction to SERS and allow the reader to make informed practical use of the method but theoretical development continues [43]. From time to time, books and symposia with many seminal articles have been published such as those in Refs. [44–46].

REFERENCES

1. Fleischman, M., Hendra, P.J., and McQuillan, A.J. (1974). *Chem. Phys. Lett.* **26**: 163.
2. Jeanmarie, D.C. and Van Duyne, R.P. (1977). *J. Electroanal. Chem.* **84**: 1.
3. Albrecht, M.G. and Creighton, J.A. (1977). *J. Am. Chem. Soc.* **99**: 5215.
4. Moskovits, M. (1985). *Rev. Mod. Phys.* **57**: 783–826.
5. Ding, S.-Y., You, E.-M., Tian, Z.-Q., and Moskovits, M. (2017). *Chem. Soc. Rev.* **46**: 4042.
6. Campion, A. and Kambhampati, P. (1988). *Chem. Soc. Rev.* **27**: 241.
7. Creighton, J.A. (1998). *Spectroscopy of Surfaces* (ed. R.J.H. Clark and R.E. Hester), 27. Wiley.
8. Otto, A., Mrozek, I., Grabhorn, H., and Akemann, W. (1992). *J. Phys. Condens. Matter* **4**: 1142.
9. Mrozek, I. and Otto, A. (1989). *Appl. Phys. A: Mater. Sci. Process.* **49**: 389.
10. Ross, M.B., Mirkin, C.A., and Schatz, G.C. (2016). *J. Phys. Chem. C* **120**: 816.
11. Schatz, G.C., Young, M.A., and Van Duyne, R.P. (2006). *Topics in Applied Physics*, vol. 103 (ed. K. Kneipp, M. Moskovits and H. Kneipp), 19–45. Berlin: Springer.
12. Faulds, K., Littleford, R., Graham, D. et al. (2004). *Anal. Chem.* **76**: 592.
13. Kleinman, S.L., Sharma, B., Blaber, M.G. et al. (2013). *J. Am. Chem. Soc.* **135**: 301.
14. Lombardi, J.R. and Birke, R.L. (2009). *Acc. Chem. Res.* **42**: 734.
15. Zhao, X. and Chen, M. (2014). *J. Raman Spectrosc.* **45**: 62.
16. Gieseking, R.L., Ratner, M.A., and Schatz, G.C. (2017). *Faraday Discuss.* **205**: 149.
17. Kneipp, K., Wang, Y., Kneipp, H. et al. (1997). *Phys. Rev. Lett.* **78**: 1667.
18. Nie, S. and Emory, S.R. (1997). *Science* **275**: 1102.
19. Cunningham, D., Littleford, R.E., Smith, W.E. et al. (2006). *Faraday Discuss.* **132**: 135.
20. Zoorob, M.E., Charlton, M.D.B., Mahnkopf, S., and Netti, C.M. (2006). *Opt. Express* **14**: 847.

21. McAnally, G.D., Everall, N.J., Chalmers, J.M., and Smith, W.E. (2003). *Appl. Spectrosc.* **57**: 44.
22. Liu, S. and Han, M.-Y. (2010). *Chem. Asian J.* **5**: 36.
23. McCabe, A.F., Eliasson, C., Prasath, R.A. et al. (2006). *Faraday Discuss.* **132**: 303.
24. Halas, N.J., Lal, S., Chang, W.-S. et al. (2011). *Chem. Rev.* **2011** (111): 3913.
25. Xie, H., Larmour, I.A., Smith, W.E. et al. (2012). *J. Phys. Chem. C* **2012** (116): 8338.
26. McLintock, A., Cunha-Matos, C.A., Zagnoni, M. et al. (2014). *ACS Nano* **8**: 8600.
27. Li, J.F., Huang, Y.F., Ding, Y. et al. (2010). *Nature* **464**: 392.
28. Li, J.-F., Zhang, Y.-J., Ding, S.-Y. et al. (2017). *Chem. Rev.* **117**: 5002.
29. Wilson, H. and Smith, W.E. (1994). *J. Raman Spectrosc.* **25**: 899.
30. Rodger, C., Dent, G., Watkinson, J., and Smith, W.E. (2000). *Appl. Spectrosc.* **54**: 1567.
31. Bersani, D., Conti, C., Matousek, P. et al. (2016). *Anal. Methods* **8**: 8395.
32. Pozzi, F. and Leona, M. *J. Raman Spectrosc.* **47**: 67. 92016.
33. McCabe, A., Smith, W.E., Thomson, G. et al. (2002). *Appl. Spectrosc.* **56**: 820.
34. McLaughlin, C., MacMillan, D., McCardle, C., and Smith, W.E. (2002). *Anal. Chem.* **74**: 3160.
35. Papadopoulou, E. and Bell, S.E.J. (2010). *J. Phys. Chem. C* **114**: 22644.
36. Papadopoulou, E. and Bell, S.E.J. (2011). *J. Phys. Chem. C* **115**: 14228.
37. Faulds, K., Jarvis, R., Smith, W.E. et al. (2008). *Analyst* **133**: 1505.
38. Gracie, K., Correa, E., Mabbott, S. et al. (2014). *Chem. Sci.* **5**: 1030–1040.
39. Lyandres, O., Shah, N.C., Yonzon, C.R. et al. (2005). *Anal. Chem.* **77**: 6134.
40. Sharma, B., Bugga, P., Madison, L.R. et al. (2016). *J. Am. Chem. Soc.* **138**: 13952.
41. de Nijs, B., Kamp, M., Szabó, I. et al. (2017). *Faraday Discuss.* **205**: 505.
42. Faulds, K., Smith, W.E., Graham, D., and Lacey, R.J. (2002). *Analyst* **127**: 282.
43. Schmidt, M.K., Esteban, R., Benz, F. et al. (2017). *Faraday Discuss.* **205**: 31.
44. Kneipp, K., Moskovits, M., and Kneipp, H. (eds.) *Topics in Applied Physics*, vol. 103, 19–45. Berlin: Springer.
45. Graham, D., Moskovits, M., and Tian, Z.-Q. (2017). Surface and tip enhanced spectroscopies themed issue. *Chem. Soc. Rev.* **46**: 3864–4110.
46. (2017). *Faraday Discuss.* **205**: 1–621.

Chapter 6

Applications

6.1 INTRODUCTION

In the previous chapters, examples of materials examined by Raman spectroscopy have been given to highlight specific aspects of the technique. However, the improvements in spectrometer design, software, data analysis and ability to overcome interference mean that the range of applications is now very large and ranges from interplanetary exploration to medical diagnostics. This chapter is designed to show how Raman scattering has been applied in some of the main areas and to provide examples which could be used to inform future developments. Some applications within the authors' experiences are given in more detail to exemplify the strengths and pitfalls of the technique which will act as a guide to the reader's potential use in their own areas of interest.

6.2 INORGANICS AND MINERALS AND ENVIRONMENTAL ANALYSIS

Raman spectroscopy has been used since its early development for the identification of inorganic materials or for structural elucidation [1–4]. It is the one of the few analytical techniques which can positively identify and characterise both elements and molecules. Unambiguous identification can be carried out for both the purity and physical form of some elements including carbon, germanium, sulphur, silicon and the halogens. For example, starting with amorphous carbon, the bands sharpen as the crystallinity increases and then reaches the ultimate with pure diamond giving a single sharp band at 1365 cm^{-1} (Figure 6.1). Spectra of amorphous carbon and graphite can also be recorded readily with visible excitation [5, 6] but with 1064 nm

Modern Raman Spectroscopy: A Practical Approach, Second Edition. Ewen Smith and Geoffrey Dent.
© 2019 John Wiley & Sons Ltd. Published 2019 by John Wiley & Sons Ltd.

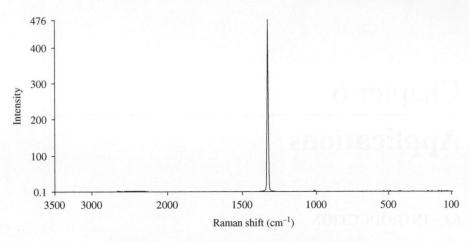

Figure 6.1.   NIR FT Raman spectrum of diamond.

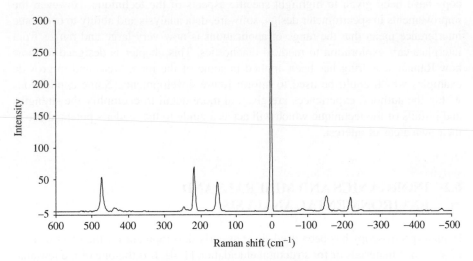

Figure 6.2.   NIR FT Raman spectrum of sulphur showing both Stokes and anti-Stokes shifts. The Stokes spectra are to the left of the exciting line at zero shift.

excitation due to the high powers often used, burning can occur unless the power and accumulation time are carefully controlled. The efficacy of Raman spectroscopy to probe the forms of applied elemental carbon is discussed in more detail in Section 6.6. Elemental sulphur is another strong Raman absorber with strong bands at ~200 cm^{-1}. It is sometimes used as a standard for instrument performance checks. The spectra do, however, vary with physical form. Flowers of sulphur (monoclinic) have a different spectrum from other sulphur forms [7]. Figure 6.2 shows both Stokes and anti-Stokes bands of sulphur.

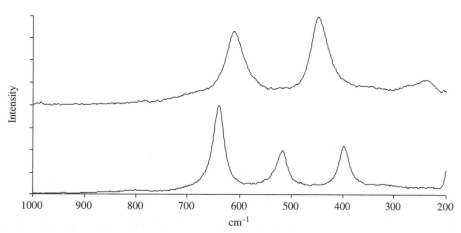

Figure 6.3. NIR FT Raman spectra of TiO$_2$ – rutile (top); anatase (bottom).

Early work carried out in Raman spectroscopy showed the strength of the technique to analyse inorganics materials. Particulates in urban dust [8] such as anhydrite, calcite, dolomite and quartz have been identified and characterised. Early microprobes [9–11] were used to record the spectra of gaseous, liquid and solid inclusions in minerals. These include $CH_4/CO_2/N_2$ ratios and solids such as apatite, calcite, nacholite and sulphur [12]. Inorganics have also been identified in biological samples, e.g. copper sulphide needles inside the lysosomes of *Littorina littorea* [13].

Figure 6.3 clearly shows the differences in the Raman spectra of the rutile and anatase forms of titanium dioxide, widely used as a white pigment and as a filler for polymers. The difference in properties is critical to use TiO$_2$ and Raman scattering has been employed quantitatively for plant control as described in Section 6.8. The breadth of the TiO$_2$ bands is normal for oxides and some other materials but many inorganic compounds have spectra with sharp bands. These sharp bands make them relatively easy to pick out in the spectra of mixtures. In many cases the Raman spectra of organic materials have strong bands in the same position as some of the bands in the infrared spectrum but with different relative intensities; however, with inorganic materials there are some notable exceptions to this. In the spectra of sulphates, the bands have very different shapes but are in similar positions in the spectrum, whilst the carbonate bands are in significantly different positions (Figure 6.4). This is due to the relative intensities of the symmetric and asymmetric vibrations.

A list of band positions observed in some common inorganic compounds is given in Appendix A. The spectra were recorded with a 1064 nm exciting wavelength. Copies of the spectra, in PDF format, are available on the internet [14]. There are also published texts [15, 16] with collections of specific Raman band positions. Using other exciting wavelengths generally does not affect the band positions but there are apparent exceptions [17]. Besides the normal background fluorescence, specific

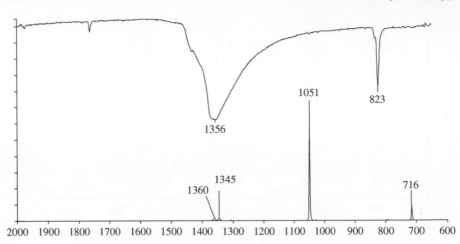

Figure 6.4. Infrared and NIR FT Raman spectra of $NaCO_3$.

sharper bands in minerals have also been reported due to fluorescence. These can be misinterpreted as Raman bands. Features to be aware of in studying the Raman spectra of inorganic materials are the changes which can occur due to the form or orientation of the crystal in the beam, as mentioned in Chapter 3. Inorganic compounds tend to be more crystalline than many organic compounds and hence more susceptible to these effects. As discussed in Chapter 2, particle size effects can also change the spectrum.

Much early Raman spectroscopy and microprobe work was carried out on minerals, for the identification of impurities and inclusions [18]. This area has found geological applications in studying both terrestrial [19, 20] and extra-terrestrial [21–23] materials. Tables of known band positions have also been published for minerals [24].

The development of handheld and standoff devices makes field geology and space exploration much more practical [23, 25–27]. The lack of sample preparation, the noncontact nature of the measurement, ease of mapping and the ability to detect and identify both organic and inorganics *in situ* are significant advantages. At the time of writing suitable rover systems for incorporation in future, Mars landers have been developed and tested on earth. The reports on tests on earth have identified different minerals and even different types of the same mineral as the results shown in Figure 6.5, which were taken in the Atacama desert, illustrate. Raman Light Detection and Ranging (LIDAR) systems [28, 29] have also been developed and used, for example, to study water, aerosols and pollution in the atmosphere.

In addition to geological samples, Raman scattering has been deployed to study pesticides, coatings and pollutants such as plastics [30, 31]. SERS can be useful here in that samples coated with a suitable substrate such as silver nanoparticles

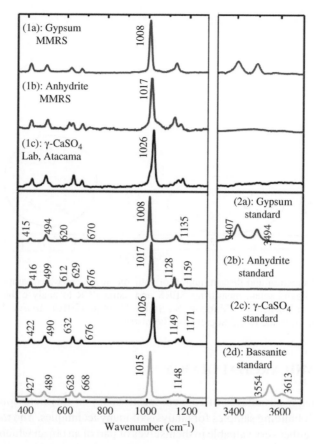

Figure 6.5. Raman spectra of three different calcium sulphates from a sample taken from the Atacama desert compared to four standards. Source: Reproduced with permission from Wei, J., Wang, A., Lambert, J.L., et al. (2015). *J. Raman Spectrosc.* **46**: 810 [26].

can detect target molecules such as pesticides directly *in situ*. Metal ions in water can be detected and quantified by reacting the ions with a ligand to form a complex which can be detected by SERS [32]. Figure 6.6 shows that excellent discrimination is obtained. In a mixture the three ions, for which WHO recommended limits were available, could be detected quantitatively well below the recommended limit.

Raman spectroscopy of inorganic compounds can be used for commercial applications such as TiO_2 plant monitoring, quantitative methods for monitoring inorganics in storage tanks [34], diamond and sapphire quality checks [35–37] and testing of various jade minerals [38]. It also has the advantage that organic species can also be picked up and identified in the spectrum, for example, carbon species in minerals have been studied [38].

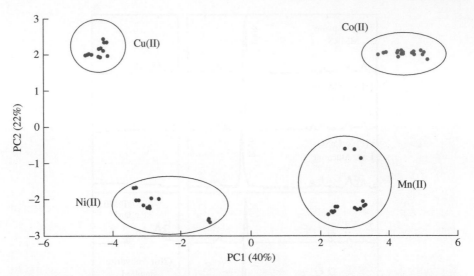

Figure 6.6. Detection and discrimination of four metal ions in water samples by reacting them with the ligand salen, measuring SERS and using PCR to analyse the data. Source: Reproduced from Docherty, J., Mabbott, S., Smith, E., et al. (2016). *Analyst* **141**: 5857 [33] with permission from the Royal Society of Chemistry.

6.3 ART AND ARCHAEOLOGY

One of the major advantages of Raman spectroscopy in this field comes from the difficulties of obtaining samples for analysis of many techniques. The materials to be examined are either very valuable in themselves or part of an object which is extremely valuable. Removing even the smallest sample for analysis would cause damage and subsequent loss of value. Raman spectra can be obtained from microsamples and, by using confocal techniques, from different layers without having to separate them. When microsamples cannot be taken, examination can take place using fibre-optic probes and/or remote sensing [39, 40]. SERS is also possible with only a microscopic amount of silver required.

There is extensive knowledge of the compositions of colour used in paintings and decorated *objects d'art*. For many centuries colouring came from inorganic pigments and natural dyes [41]. There were very few synthetic dyes available until the nineteenth century [42]. Raman spectroscopy can not only identify the type of inorganic materials used but also the physical forms. The use of the dyes, pigments and resins [43] can be established chronologically [44]. By examining the Raman spectra of paintings [41, 45–48] and archaeological artefacts such as pottery [49] the age of the work can be determined. Figure 6.7 demonstrates this feature very well. The knowledge of the composition can be used to distinguish original work from restoration and/or forgery [51].

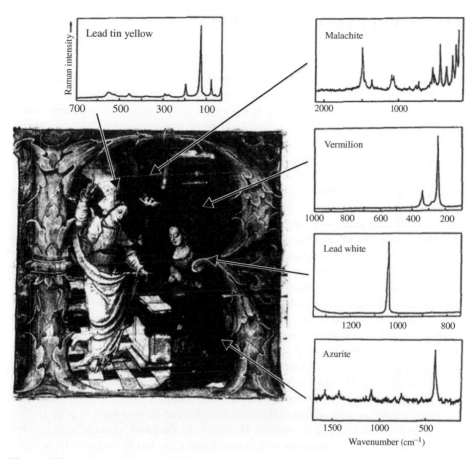

Figure 6.7. Spectra from a sixteenth-century German choir book: historiated letter 'R'. Source: Reproduced from Clark, R.J.H. (1995). *J. Mol. Struct.* **347**: 417–428 [50] with permission from Elsevier.

Besides colour identification, the identification of gemstones [52, 53], porcelains [54], metal corrosion products [55] and organic materials such as resins [56] and ivory [57] makes Raman spectroscopy a very valuable technique in this field. Ivory, of particular interest, has been studied for environmental purposes and by law enforcement bodies. Vibrational spectroscopic assignments of mammalian ivories have been published [58]. Both the art and archaeological areas show growing applications of Raman spectroscopy for historical research, authentication and forgery.

6.4 POLYMERS AND EMULSIONS

6.4.1 Overview

The applications of Raman spectroscopy to polymers are extensively reported in the literature. One publication [59] devoted 10 chapters to the vibrational analysis of polymers with Raman spectroscopy being extensively used. Only a general overview is given here in an attempt to highlight some of the strengths. Polymers have been studied for identification, structure, composition, cure and degree of polymerisation in the solid, melt, film and emulsion states. Ironically, polymers, particularly aliphatic ones, are not very strong Raman scatterers and sample preparation techniques such as folding thin films have to be employed. Conversely, items such as aspirin can be studied inside a film wrap without interference from the film. In the early days Raman studies of some polymers were restricted by fluorescence and thermal absorption due to impurities and fillers. This problem has been overcome to a great extent by the increased ability to choose an appropriate excitation frequency for specific samples anywhere from the infrared to the UV. Cleaner processes leading to fewer residues have also helped. However, some commercial products still contain antioxidants, plasticisers and fillers which can cause interferences. Both the chemical and the physical nature of the polymers themselves have to be considered before examination. Whilst Raman spectroscopy can be considered a minimal sample preparation technique, the physical state (granules, film, etc.), the morphology (macro- and microcrystallinity), thermal properties (high/low melting), state of cure, copolymer distribution and homogeneity of fillers does affect the way in which a sample is best presented to the Raman spectrometer. Many samples can be placed directly in the beam for 90° or 180° signal collection. Samples can be presented in glass bottles and aqueous emulsions can be studied. However, for the latter 'particle' size relative to the exciting wavelength needs to be considered. Care has to be taken if, in order to detect weak scatters, the laser power is increased. This may cause thermal damage or induce changes.

6.4.2 Simple Qualitative Polymer Studies

Of the many in-depth and wide-ranging reviews written on the Raman spectroscopy of polymers, a few relatively simple applications are described here to illustrate the range of the technique. A published collection of spectra of common polymers is available [60, 61]. The spectra of five of the most commonly encountered polymers, polyethylene (PE), polypropylene (PP), polyethylene terephthalate (PET), polycarbonate and polystyrene were all recorded quickly and easily with no sample preparation using an FT system. In Figure 6.8 the spectra are quite distinctive even showing subtleties of chain branching between PE and PP in the 2900 and 1450 cm^{-1} region of the spectrum. Figure 6.9 shows that whilst the polycarbonate spectra are dominated by the bands due to the aromatic groups, differences due to the aliphatic methyl and cyclohexyl groups are clear. The bands are in the same spectral regions

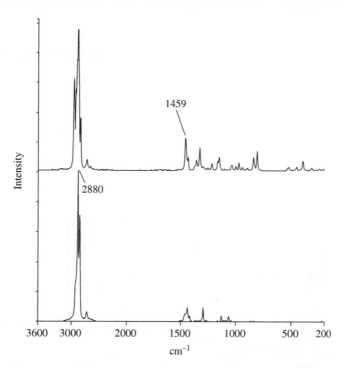

Figure 6.8. NIR FT Raman spectra of polypropylene (top); polyethylene (bottom).

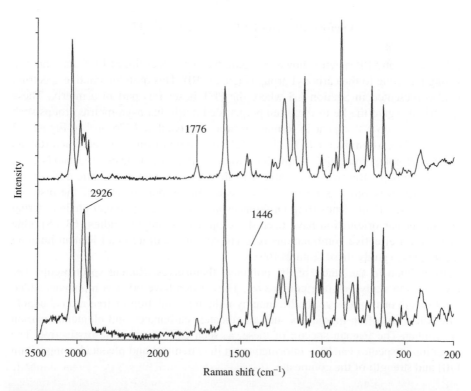

Figure 6.9. NIR FT Raman spectra of two polycarbonates with different aliphatic methyl and cyclohexyl groups.

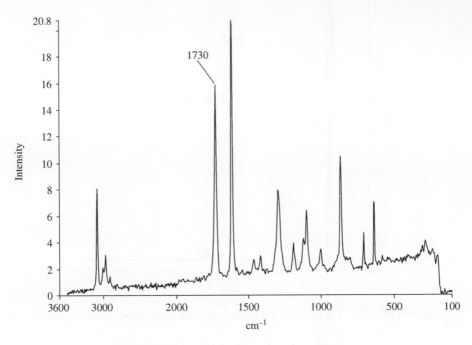

Figure 6.10. NIR FT Raman spectrum of PET.

as in the PE and PP spectra shown in Figure 6.8. The spectrum of PET also shows a strong band due to the carbonyl group (Figure 6.10). This quite distinctive spectrum is also presented in Section 6.5 where the PET is seen as part of a matrix. These spectra also give the lie to the often perpetrated myth that asymmetric groups such as carbonyls do not appear in Raman spectra. The band at $1776\,cm^{-1}$ in Figure 6.7 is due to the carbonyl stretch. In Figure 6.11 the comparison of infrared and Raman spectra of polystyrene does show how Raman spectra emphasise the bands from aromatic groups. Because of the ability to show these sometimes subtle differences, Raman spectroscopy has been used to create microscopic images of the distribution of polymers in blends [62]. Differences in morphology, polymer chain ordering and molecular orientation have been the subject of in-depth studies [63–68]. One of the most effective combinations is to employ both infrared and Raman imaging techniques to study these features [69].

In addition to the study of the polymers themselves, Raman spectroscopy has been used to study polymer composites. These often have other components added for strength or to preserve the lifetime by reducing oxidation or free radical attack. The fillers are often inorganics such as silicates, carbonates and elemental carbon or sulphur. The Raman spectra of these, as seen earlier, are quite distinctive. The variation in spectra can give information on the chemical and physical composition [70] and strength of the composite.

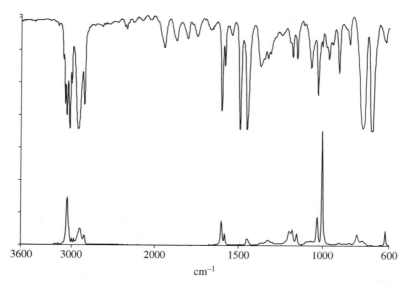

Figure 6.11. Spectra of polystyrene – infrared transmission spectrum (top); FT Raman (bottom).

Raman spectra can identify the type of polymer present in a material but can also be used to study how far polymerisation has occurred [71, 72] or even how far a polymer has been degraded [73, 74]. The double bond in acrylates is very strong and quite characteristic. As polymerisation occurs, the strength of this bond decreases. This can be easily monitored by Raman spectroscopy and has been developed for a number of applications for plant scale monitoring. Acrylate-based emulsions are quite common. In another example, the variation in the band due to >C=C< at 1655 cm⁻¹ can be seen in Figure 6.12 for sunflower oil. The top spectrum is an emulsion with degraded sunflower oil. The bottom is a Raman spectrum of the pure oil.

Following the polymerisation of dispersions of polymer in aqueous media is very difficult by conventional methods. However, by using a fibre-optic probe, following the reduction of a band from the monomer by Raman scattering is quite feasible, in glass vessels under manufacturing conditions. Whilst this experiment is relatively easy to carry out in principle, one must consider the relative size of the emulsion drops in the suspension and the laser wavelength employed. This is similar to the particle size effects discussed in Chapter 2. The opposite of following a polymerisation reaction is to monitor degradation of a polymer. One of the earliest reported degradation studies [75] was on polyvinyl chloride (PVC). As PVC degrades, hydrogen chloride (HCl) is lost and conjugated double bonds form. The wavenumber position in the Raman spectrum shows the number of bonds in a chain which have been formed and hence the degree of conjugation and degradation. The studies by Gerrard and Maddams showed intense bands at 1511 and 1124 cm⁻¹ associated with conjugation

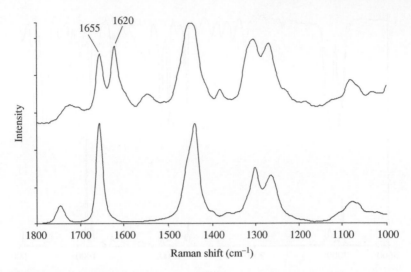

Figure 6.12. NIR FT Raman spectra of oils – emulsion (top); pure oil (bottom).

unsaturation. By variation of the illuminating laser line they showed that the intensity was due to resonance effects. They also showed that the position of the band due to >C=C< between 1650 and 1500 cm^{-1} correlated with the length of the sequence. Subsequent workers [76–78] have used these bands to study similar sequences in various polymers.

6.4.3 Quantitative Polymer Studies

The applications described have largely dealt with the physical and chemical characterisation of polymers. In several of the cases cited, quantitative aspects can also be measured. Quantitative measurements in Raman spectroscopy of polymers vary from the relatively simple to the quite complex. The relative intensities within a normal Raman spectrum will simplistically be directly proportional to concentrations of the species present, the laser power and the Raman scattering cross-section. As the scattering cross-section is very difficult to determine, absolute band strengths are rarely, if ever, determined. Determination of relative strengths by using band ratios is most common. This method can easily be employed in the examples already cited of PVC degradation, degrees of polymerisation of acrylics and epoxides and filler content. More complex studies often have to employ more sophisticated quantitative techniques which involve several bands or complete regions of the spectrum. This is particularly the case where multicomponent blends, composites or morphological features are being studied. An example of this is the modelling of PET density from normalised, mean-centred FT Raman spectroscopy studies. The bandwidth of the carbonyl band in the Raman spectrum has been associated with the density and hence the sample crystallinity [79].

6.5 DYES AND PIGMENTS

6.5.1 Raman Colour Probes

Raman spectroscopy can be a very sensitive probe of coloured molecules particularly in resonant conditions (Chapter 4). Resonance increases the intensity of selected vibrations giving in many cases a limited set of sharp and intense bands so that the coloured species can be identified *in situ* even as a minor component in a mixture. However, as discussed in Chapter 4, coloured molecules are prone to thermal degradation. There are ways of diminishing this effect by spinning solid samples, flowing liquids or diluting solids in a matrix such as hydrocarbon oil or potassium bromide (KBr) powder. For some samples, fluorescence can also be a major problem. Moving to a higher wavelength can sometimes mitigate these problems but even small amounts of some colours in a sample can still cause problems. Moving to an infrared excitation frequency will overcome these problems in most but not all cases but at the expense of all or most of the resonance enhancement. SERS and SERRS described in Chapter 5 can both overcome the fluorescence effects and provide a large increase in scattering. For suitable molecules, SERS gives an increase in sensitivity of about 10^6 or more and also strongly quenches fluorescence and resonance Raman scattering often gives an increase of 10^3–10^4. The combination of the two techniques, surface enhanced resonance Raman spectroscopy (SERRS), gives a greater enhancement. Rhodamine dyes have proved to be very effective for SERRS. Early SERRS studies of rhodamines 3G, 6G (Figure 6.13) and 3B (**1**) have detected the dye in solution at less than 10^{-17} mol l^{-1} and more recent studies have claimed detection limits of ~10^{-18} mol, which is roughly equivalent to having 35 molecules in the beam at a given time [80–83].

(1)
Generic structure of
rhodamines

These low detection limits have opened up previously prohibited areas of vibrational spectroscopy. This is particularly true in the biological area. Infrared spectroscopy is limited by the strong absorption of water but conversely, Raman spectroscopy can cope with aqueous media and the applications are expanding. One area of much interest is tagging DNA with a chromophore to obtain SERRS. The chromophore can be a fluorophore since fluorescence quenching will remove

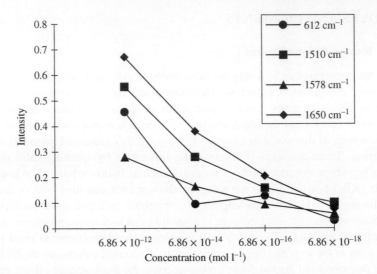

Figure 6.13. Graphical representation of intensity vs. concentration for four peaks selected from the R 6G SERRS spectra using nitric acid aggregation (points corresponding to R 6G concentration of $6:8 \times 10^{-10}$ M were omitted). Source: Reproduced from Rodger, C., Smith, W.E., Dent, G., and Edmondson, M. (1996). *J. Chem. Soc. Dalton.* **5**: 791 [80] with permission from the Royal Society of Chemistry.

the emission from molecules on the active surface (see Section 5.7). Nonfluorescent chromophores are also effective. Using SERRS, levels of DNA at 10^{-15} molar have been detected [84–86] and the results from SERRS experiments using these tags are shown in Chapters 5 and 7.

6.5.2   *In Situ* Analysis

One major advantage of Raman spectroscopy with coloured molecules is that vibrational spectroscopy can be used as a largely nondestructive, *in situ* sampling technique to directly study these molecules in the applications for which they were designed. The amount used is often relatively low, e.g. a 2% dye loading on a polymer, but this can still be detected effectively by Raman spectroscopy [87]. The low concentration has the effect of diluting the colour and reducing thermal degradation, similar to the mull and halide disk techniques. The strong bands which appear in the Raman spectrum are from the part of the molecule generating the chromophore and weak bands appear from the rest of the molecule. The reason for these extra-strong bands with visible laser sources is due to resonance enhancement. However, in some cases bands from the chromophore are still the strongest with 1064 nm excitation even for dyes with only weak or no absorption band in this region. The likelihood is that this is due to the normal strong scattering (high cross-section) which can arise from the chromophore structure which often contains aromatic rings and also from some pre-resonance [88] (see Section 4.3.1). This feature can be used to study changes in dye conformers and the effect or otherwise on the chromophore not only on the neat dye but also that dispersed in the material being dyed. A disperse red dye (**2**) for polymer textiles has been shown to exhibit more than one physical form by liquid and solid NMR studies [89, 90]. NIR FT Raman spectra of the solid samples of the dye showed differences in the azo band positions and also in the backbone structure of the molecules. A piece of PET ester cloth dyed at ~2% level with disperse red was examined directly in the spectrometer beam with 1064 nm excitation. The spectra of forms I and II are shown in Figure 6.14a. The dyed cloth and undyed cloth are shown in Figure 6.14b.

The bottom spectrum of Figure 6.14b whilst showing strong bands due to the PET fibres also shows bands due to the dye. It is clear from this spectrum that form I of the dye is predominant whilst the dye is in the fibre.

(**2**)
Disperse red

SERRS also enables *in situ* identification of chromophores from small samples. This technique has been used to identify chromophores in pen inks [91], dyed fibres [92, 93] and lipsticks [94]. With polymer fibres the dyes are dispersed in the fibre but with cellulose the dyes can be reacted onto the fibre. The SERRS technique has been used to detect dyes, at very low levels, reacted onto the fibre [68] as shown in Figure 6.15. The lower spectrum shows the dye in solution. The upper spectrum is of the dye attached to a fibre which had been treated in caustic solution for two hours. This was to ensure that the dye was firmly attached and would not leech back into the SERRS colloid.

Dyes and pigments are used for printing in a number of different printing mechanisms and applications [95]. One of the more commonly encountered uses of dyes is in the ink of inkjet printers. The dyes first developed for these inks were very similar to those used for textiles. The requirements of the ink manufacturers were very different from the textile dyers. Only three main colours, cyan, yellow and magenta, were required. The dyes had to have high colour, with good light and wet fastness. Unlike textiles the dyes could not be fixed to the paper by boiling! Dye development occurred along the non-chromophoric part of the dye to increase fastness to the inkjet media. Originally, the inkjet medium was cellulose paper of varying complex composition and pH, but now it can be polymers, gels, textiles or electronic surfaces. Total internal reflection Raman (TIR) has been employed to study the distribution of dyes in coated paper surfaces [96]. Studies of the behaviour of the dyes

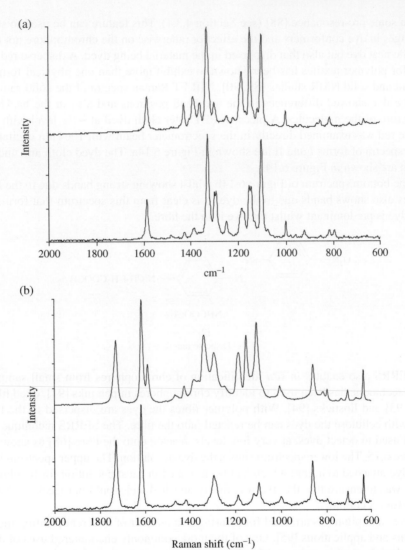

Figure 6.14. NIR FT Raman spectra of (a) azo dye, forms I (top) and II (foot); (b) red cloth – dyed (top); dye in PET (bottom).

on various types of paper surfaces to determine the effects on the chromophore of pH have been carried out [97, 98]. By using both SERRS and NIR FT Raman techniques, the behaviour of the dyes can be studied as solids, on the surface of the paper and below the surface. The effect of varying the non-chromophoric constituents on the chromophore in various parts of the media under differing pH conditions can be studied in this way.

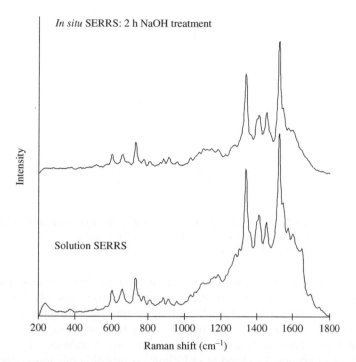

Figure 6.15. SERRS spectra of reactive dye attached to a cellulose fibre and in solution. Source: Reproduced from White, P.C., Rodger, C., Rutherford, V., et al. (1998). *SPIE* **3578**: 77 [91] with permission from the Royal Society of Chemistry

6.5.3 Raman Studies of Tautomerism in Azo Dyes

The changes seen in the *in situ* analysis of dyes can be due to physical or chemical form changes. Azo dyes are amongst the most extensively employed dyes in the colour industry. They are used for their electrical properties as well as their wide colour range. The azo group (**3**), which simplistically exhibits azo hydrazo tautomerism, is symmetrical and therefore very weak in the infrared spectrum. The band can be strong in Raman spectra [99] at ~1450 cm^{-1}. When the hydrazo is formed, the —C=N— band can be seen in the Raman spectrum at ~1605 cm^{-1} with another strong band at ~1380 cm^{-1}, the origin of which is still not fully understood. Whilst this group is very important in colour chemistry, being a significant component in many dyes [100–102], the detection and interpretation of these bands can be complex and have been the subject of many studies by the authors.

(3)
Azo-hydrazo tautomers

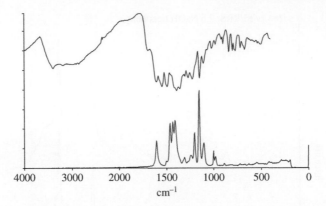

Figure 6.16. Infrared and NIR FT Raman spectrum of an azo dye. Source: Reproduced with permission from Chalmers, J. and Griffiths, P. (eds.) (2001). *Handbook of Vibrational Spectroscopy*, vol. 4. New York: John Wiley & Sons, Inc. [59].

A simple example of the comparative information obtained from the infrared and Raman spectra is shown in Figure 6.16 for a yellow diazo dye (**4**). The dye has a generic structure with azo, carboxyl and triazine groups. The upper infrared spectrum is complex due to strong hydrogen bonding between groups. Bands between 3500–2000 and 1700–1500 cm^{-1} are consistent with a mixture of salt and free carboxylic acid groups. The band at ~1550 cm^{-1} is probably due to the triazine ring. The general broadness is due to both the hydrogen-bonding effects and the large size of the molecule. The azo groups cannot be seen in this infrared spectrum.

a(HOOC)

a = 1 or 2
b = 1 or 2
X= morpholinyl or NHR

(**4**)
Generic structure of yellow diazo

Conversely, in the lower Raman spectrum the azo and aromatic bands dominate the spectrum, with the hydrogen-bonded groups being too weak to observe.

6.5.4   Polymorphism in Dyes

In the azo dyes, chemical group changes, which could affect the colour properties, were studied. Physical changes can also take place in the molecular structure of a dye which in turn can also affect the properties. This effect known as polymorphism is described in Section 6.7 and an application of the effects is also shown in Section 6.6.

6.6 ELECTRONICS APPLICATIONS

In electrophotography, the selectivity of resonance Raman scattering make it a good probe for dyes including phthalocyanine dyes used in photoreceptors. For example, in a photocopier or laser printer, one stage of copying the image requires a charge to be generated by excitation of a photoreceptor with a laser. The film containing the receptor usually comprises several layers. The base is usually a conductor, above which is a charge transfer generation (CTG) layer containing the photoreceptor. Upon exposure to the laser, an electronic hole is generated. A charge is transferred to the surface of the film through a charge transport material (CTM) in a charge transport layer. The charged surface is then used to create the image. The CTG is less than 1 μm thick and dyes are often used to absorb the laser energy and create the hole.

(5)
Metal-free phthalocyanine

Phthalocyanines (**5**), which give excellent resonance Raman spectra, are often used as the dye. Their electrical properties vary with the metal coordinated in the ring where the metal size can either fit the hole, cause buckling of the ring or sit in a position slightly above the ring. The positions of the ring breathing band at ~1540–1510 cm^{-1} and other bands in the Raman spectrum are quite characteristic of each phthalocyanine as can be seen in Figure 6.17a and the spectra are intense enabling *in situ* identification and distribution.

Titanium phthalocyanine (TiOPc) has several polymorphs and one, Type IV, is widely used since it is gives optimum charge generation. The Raman spectra of various TiOPc polymorphs in Figure 6.17b show subtle but distinctive changes enabling Type IV, which has a very small extra band at 765 cm^{-1} to be identified *in situ*. The spectrum of a drum placed directly in the beam of an NIR FT Raman spectrometer (Figure 6.18) shows this. Bands are present from other components which have high Raman cross-sections and are present in bulk amounts. In particular, the CTM tends to be based on triarylamines which have strong aromatic bands due to the multiple conjugation and the base layer of titanium dioxide which in Figure 6.18 is present as anatase (see Chapter 2).

A similar approach is taken to the analysis of organic semiconductors [104] such as field effect transistors [105–107], solar cells [108] and light-emitting displays [109, 110]. The types of materials employed in these devices are usually conducting

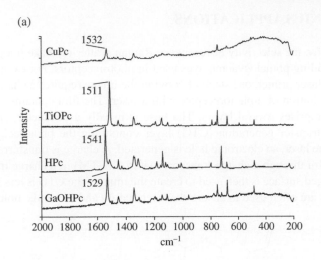

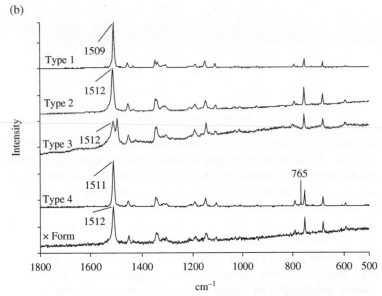

Figure 6.17. (a) NIR FT Raman spectra of various metal phthalocyanines. Source: Reproduced from NIR FT Raman examination phthalocyanines at 1064 nm, Dent, G. and Farrell, F. (1997). *Spectrochim. Acta* **53A** (1): 21 [103] © 1997 by kind permission of Elsevier Science-NL, Sara Burgerhartstraat 25, 1055 KV Amsterdam, The Netherlands. (b) NIR FT Raman spectra of the polymorphs of TiOPc. Source: Reproduced from NIR FT Raman examination phthalocyanines at 1064 nm, Dent, G. and Farrell, F. (1997). *Spectrochim. Acta* **53A** (1): 21 [103] by kind permission of Elsevier Science-NL, Sara Burgerhartstraat 25, 1055 KV Amsterdam, The Netherlands.

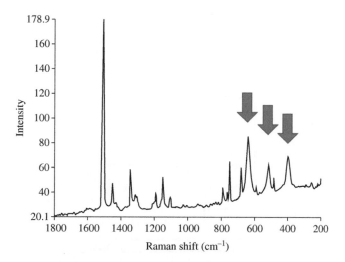

Figure 6.18. NIR FT Raman spectra of photocopier drum showing the spectrum from phthalocyanine and, indicated by arrows, TiO₂ in the form of anatase. Source: Reproduced from Chalmers, J. and Griffiths, P. (eds.) (2001). *Handbook of Vibrational Spectroscopy*, vol. 4. New York: John Wiley & Sons, Inc. [59].

polymers with conjugated π-electrons. Typical polymers initially employed were poly-acetylene, poly(*p*-phenylenes), polythiophenes and poly(triarylamines) [111–114]. Several of these have been available from commercial sources since 1996 [115]. Poly (*p*-phenylene vinylene)s (PVPPs) and spiro compounds have also been developed [116, 117]. A typical band gap of these polymers is 1–3 eV. A conjugated polymer can be doped with electron acceptors such as halogens, or donors such as alkali metals. The polymers can be discussed in concepts familiar to physicists of solitons [118], polarons [119] and bipolarons [120]. Essentially the polymers carry charges which can move through the material and may, for example, emit light. As in other polymer applications the effectiveness of these materials is dependent on the morphology as well as the chemistry. The electronic behaviour of the building block monomers, on which the polymers are based, can be studied by cyclic voltammetry (CV) and the Raman spectra of the materials can be obtained during the voltage cycle to track changes.

Elements such as carbon, germanium and silicon are widely used in the electronics industries for both their mechanical and electronic properties. This area is one of the largest in which Raman spectroscopy is used as a quality control (QC) tool. Figure 6.1 shows the spectrum of diamond making it relatively straightforward to study diamond films [121] on metals such as silicon, but many forms of carbon can be easily distinguished as shown in Figure 6.19 [123]. This figure does not show the absolute intensities of different forms. Most carbon species absorb in the visible region and the scattering is resonantly enhanced. As a result, the intensity of many carbon species is much greater than for diamond and overtone and combination bands

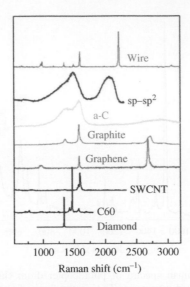

Figure 6.19. Spectra of different forms of carbon. a-C is amorphous carbon. Source: Reproduced from Milani, A., Tommasini, M., Russo, V., et al. (2015). *Beilstein J. Nanotechnol.* **6**: 480 [122] with permission from Beilstein.

occur. With visible excitation, amorphous carbon species [122–124] give two bands due to sp^2 hybridised carbon atoms at about 1360 and 1500 cm^{-1} (D and G bands). The reason for this is that π bonds are formed which absorb in the visible region and there is resonance enhancement. Bands in graphite and graphene are also due to sp^2 hybridisation but in this case in the form of ordered arrays of rings which give much sharper spectra [125, 126]. The band which occurs in both graphite and graphene at about 2600 cm^{-1} is an overtone of the D band (2D). Going from single layers of graphene to the three-dimensional structure of graphite changes the electronic struc- ture from two dimensions to three dimensions and the way the layers pack affect the Raman scattering making it a very sensitive probe of structure. Diamond shows bands which arise from sp^3 hybridised carbon which are much stronger with UV excitation but quickly disappear with visible excitation in mixtures of amorphous carbon and diamond as the diamond content is reduced. Sp hybrids also form, for example, in carbon 'wires' consisting of a single chain of carbon atoms, again with resonance enhancement with visible excitation [122]. There are two forms of these wires, sp^2 hybridised linear chains of ethene groups and sp hybridised chains of more local- ised ethyne groups. These can be distinguished by Raman scattering. Other related systems such as single walled carbon nanotubes (SWCNT) for which a spectrum is shown in the figure also give distinctive spectra.

   More detailed analysis of vibrations in these systems needs to take into account the extended nature of many of the entities present such as the flat sheets in graphene where there is no single definable molecule except the whole sheet. This requires an interpretation in terms of lattice modes and scattering from Brillouin zones. The concept of lattice modes was introduced for a simple case in Chapter 3. As a reminder, in crystals, as well as intramolecular bonding in molecules, there are longer range interactions between ions or molecules and as a result, on excitation, lattice modes which extend over many ionic or molecular sites are formed. The direction of the displacements of these modes is defined with reference to the incoming radiation. A longitudinal optic mode LO travels along the direction of the incoming radiation and neighbouring entities move in opposite directions and in a longitudinal acoustic (LA) mode the neighbouring entities move in the same direction. Transverse optic (TO) modes are doubly degenerate and spread out at right angles to the propagation direction and again neighbouring entities move in opposite directions and transverse acoustic (TA) modes travel in the same direction but with the neighbouring entities moving in the same direction. Clearly in carbon systems where coupled sheets of rings and other larger systems appear, there can be many lattice modes and because the interactions are strong they can occur at higher frequency. Brillouin zones, a description of the electronic states of a system based on reciprocal space, lie beyond the scope of this book, but it is necessary for an in-depth understanding. This is because the bands arise from resonant enhancement and the electronic structure contains delocalised electrons throughout the material making band theory applicable to understand the electronic properties essential for resonance. An analysis of these modes for graphene is given in Ref. [125]. However, as the work described above shows, the changes observed experimentally can still be used to investigate structural variation simply and effectively.

In addition, where an impurity is added such as a dopant in silicon, the impurity disrupts the lattice and can form a local mode which will be sharper and at a different frequency [127, 128]. Figure 6.20 shows the spectra of the LO mode from zinc selenide with different levels of an isotope of selenium substituted into the matrix [128].

Raman maps can show stress patterns and crystallinity in silicon wafers [130, 131]. In heavily boron-doped silicon the LO mode, which is a very sensitive Raman band to the environment, shifts to lower frequency and broadens as the boron concentration increases [132]. Films of fluorine-doped silicon dioxide have been monitored by Raman spectroscopy to determine the fluorine to oxygen ratios and hence the dielectric properties [133]. Silicon crystals are employed in ultra large-scale integrated (ULSI) circuits [134]. Wide-band semiconductors also give informative Raman spectra [135]. The Raman spectra can give information on the lattice modes and hence changes in the crystals. The structural characterisation can include the crystallinity, crystallographic orientation, superlattices of mixed crystals, defects and stacking faults. Besides the structural characterisation, electronic characterisation can be carried out. Both bound and free charges can contribute to Raman scattering, through collective and single-particle excitation processes.

Lattice modes are of value in understanding materials' structure over a much wider range of materials science. They appear quite strongly in three-dimensional structures such as oxides, nitrides, sulphides and selenides. For example, much of the same information as already discussed can be obtained for gallium nitride used to make

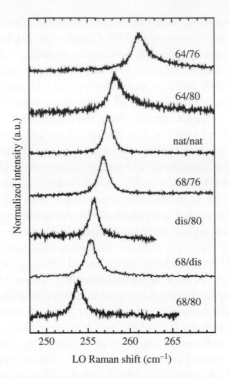

Figure 6.20. LO mode of zinc selenide with different selenium isotope ratios substituted into it. Source: Reproduced from Gobel, A., Ruf, T., Fischer, T.A., et al. (1999). *Phys. Rev. B* **59**: 12612 [129] with permission from the American Physical Society.

light-emitting devices [135]. Ref. [135] also discusses the effect of the direction of the exciting radiation in non-cubic lattices. Iron oxides are a good example. In addition, in insulators of this type unpaired electrons from the iron can interact to excite cooperative movement of the electrons. This can begin with a change of spin at one site caused by excitation of the material which travels through the material causing a spin wave called a magnon. This wave can show up in Raman scattering. A typical example is in iron oxides often used as pigments and for which an extra peak due to this wave may require to be assigned. This can cause confusion in assigning the spectrum.

6.7 BIOLOGICAL AND CLINICAL APPLICATIONS

6.7.1 Introduction

The advantages of standoff detection, *in situ* molecule identification and ability to obtain signals in aqueous environments make Raman spectroscopy very useful for analysis of biological samples. Carbonyls, amines and amides give weaker bands in

Raman scattering than in IR absorption but are still easily detectible and, in addition, groups such as —S—S—, —SH, —CN, —C≡C—, aromatic rings and carbonates and phosphates give distinctive bands. Thus, there is a wealth of molecularly specific data in biological samples. The problem is that the number of groups giving specific vibrations is often large and since they are often in slightly different environments this can lead to spectra with broad bands and ill-defined structure. However, the molecularly specific information is still there and modern data-processing methods such as PCA or more advanced methods can often be used to extract information from such samples even if the result is not clear to the eye [136]. Raman scattering cross-section varies widely between specific moieties so that a particular species with a high cross-section can be picked out directly in the mixture. This can be very successful for detecting such species as drugs or metabolites, if the molecule of interest is a strong Raman scatterer. It has also enabled the identification of cells, spores, bacteria, etc. by their specific signatures even if the individual molecules which give the signature are not always clearly identified. In some cases, in addition to identification, structural information can be obtained as discussed for the heme group in Chapter 4 and for the secondary structure of proteins in Chapter 7. Individual DNA bases can be identified in complete DNA sequences and quantified. However, the secondary structure influences the signal strength, so care is required with this method [137]. Alternatively, labels can be very effective in the detection of specific molecules in a complex matrix. These can be natural labels such as a heme group or flavin or added ones such as dyes and other tags or labelled metal nanoparticles for SERS. There is a review of a wide range of tags for cells [138]. Labels have also been used to enable the selective detection of DNA fragments or proteins in mixtures, for example, in the SERS studies described in Chapter 5.

The technique can also detect changes in physical form, so polymorphism, secondary structure in peptides and general molecular backbone changes can be detected. Among a vast number of applications reported are binding studies, genomics, proteomics, protein interactions and solid-phase synthesis. DNA and protein arrays have been developed and food analysis and cell growth studied. Early examples include studies on transitions in amino acid crystals [139], single-cell bacteria [140], bacterial spores [141], carotenoids in numerous systems including Atlantic salmon [142], characterisation of microorganisms [143], fungi [144], grain composition [145], liposome complexes [146] and yeast [147]. More recently, reviews of areas such as lipids [148], live cells [149, 150] and regenerative medicine [151] can be found which summarise particular areas.

Greater understanding and technique advances have brought the use of Raman spectroscopy in medicine much closer [152]. SORS and transmission Raman have been used to examine and define diseased tissue at depths well below the surface [153]. An early development was the design of a probe to look for Barrett's oesophageal cancer *in situ* in the throat [154] and now a range of probes have been developed for different targets. Critical to the use of these probes are the improved sensitivity and speed achieved by using mapping and scanning methods and advanced software. More advanced spectroscopy methods described in Chapter 7 are also used. One aim is to differentiate diseased tissue from healthy tissue. For example, when a cancer is removed, it is important to remove all cancer cells from the surrounding area without unnecessarily removing healthy tissue. Using probes and advanced data-processing methods to pick out the key information, Raman scattering can discriminate between healthy and cancerous tissue *in situ* without the need for biopsy during an operation. Successful studies for a wide range of cancers have been reported [154] as have initial studies of its use to detect infiltrating cancer cells in tissue surrounding brain tumours [155].

To help probe deeper into tissue, SORS has been combined with SERS and signals have been reported from particles 40 to 50 mm deep. Working in the infrared should help this since greater tissue penetration is possible. Chalcogenpyrylium dyes, which are resonant in the near infrared, have been developed and used to label gold nanoparticles so that molecularly resonant SESORS (SESORRS) can be obtained. Figure 6.21 shows a SESORRS map of these nanoparticles incorporated into multicellular tumour spheroids, an ex vivo breast cancer model cell cluster, recorded through 15 mm of pork fat. A handheld instrument was used in backscattering mode. The position of the spheroids can be seen in the SESORRS map and the spectrum in the area where they are shows peaks from the labelled nanoparticles.

6.8 PHARMACEUTICALS

The advantages of Raman spectroscopy to the pharmaceutical community come largely from the ease of use, minimal sample handling and strong differences in relative scattering strengths of packaging materials, tablet excipients and the active agents. These strengths combined with the use of microscopes and fibre optics have seen a large growth of use in the pharmaceutical industry. An early worker with FT Raman quickly recognised the advantages and opportunities in the pharmaceutical industry [157]. In the area of fibre-optic coupling, microprobes and imaging, dispersive technology is very much at the forefront. More recently, the use of SORS [158, 159] and transmission Raman spectroscopy [160–162] has become more widely used. However, as with confocal microscopy there are some issues around the beam sampling position [163].

For QC of manufacturing and formulation, the ability to check directly inside a polymer package produces tremendous time and cost savings. Imaging of tablets can be carried out to check the distribution and relative amounts of active agent, additives and binders present. The active drug is often an aromatic-based compound with distinctive Raman spectra whilst the other components are sugar, cellulose or inorganic-based materials. The active component itself can also have variable properties dependent on the physical form or crystallinity. These can affect dissolution rates and hence the efficacy of the drug. Drug samples, including both prescription drugs and drugs of abuse, can be measured *in situ* in clear plastic wrappings. This can be important both in the speed of analysis and in preventing sample contamination. If a microscope system is used, the laser beam can be focused onto the surface of

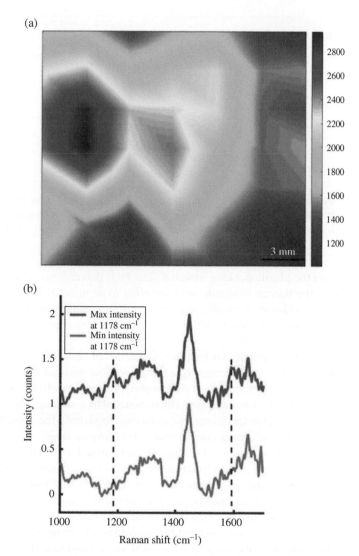

Figure 6.21. SESORRS from labelled nanoparticles incorporated into multicellular tumour spheroids, an ex vivo breast cancer model cell cluster, and recorded through 15 mm of pork fat using 830 nm excitation. The data were obtained with a handheld spectrometer clamped to enable mapping (a) shows a map of SESORRS intensity showing where the spheroids containing the nanoparticles are located and (b) signals from the dye from the area containing the spheroids (top) and the area containing few spheroids (foot). The dye peaks are on the dotted lines. Source: Reproduced from Nicolson, F., Jamieson, L.E., Mabbott, S., et al. (2018). *Chem. Sci.* **9**: 3788 [156] with permission from the Royal Society of Chemistry.

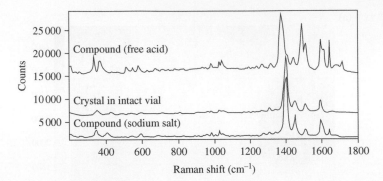

Figure 6.22. Noninvasive Raman microscopy analysis of a crystal in an intact vial of intravenous formulation. Source: Reproduced with permission from Chalmers, J. and Griffiths, P. (eds.) (2001). *Handbook of Vibrational Spectroscopy*, vol. 5. New York: John Wiley & Sons, Inc. [164].

the tablet through the plastic packing material. The high-power-density area created provides most of the Raman scattering and therefore discriminates in favour of the tablet or powder. In addition, the scattering from the drug is usually relatively intense compared to that from the plastic material. An example of this type of experiment is shown in Chapter 1, Figure 1.10.

Raman scattering can be obtained from a set of drugs of abuse. Each spectrum is molecularly specific. Initial identification of a sample without the matrix is very simple by Raman scattering. However, in real-world samples the drugs are often in a matrix of several compounds. This matrix can cause fluorescence which swamps the image; in the case of inorganics, Raman spectroscopy can be used to identify the impurity.

Similarly, the stability of drugs can be monitored directly inside the container. Figure 6.22 shows the Raman spectra of crystals that formed in a vial containing a drug/saline solution, a common formulation in early stages of drug development. Identification of these crystals was important, but the amount of sample available was very limited. Using Raman microscopy, the sample was studied noninvasively. The crystals were viewed directly through the glass vial using the microscope and Raman spectra were acquired with no need for further sample preparation. The spectra identified the crystals as the drug in its sodium salt form rather than as its free acid form. This observation led to a reassessment of the salt's solubility profile and a modification to the formulation's composition.

Crystalline active agents can have a property, polymorphism, which can greatly affect the efficacy of a drug. The term can be used in the biological and pharmaceutical worlds with totally different meanings. In most of the chemistry applications and here in pharmaceutical applications, McCrone's definition [165] is used to mean differing physical forms of the same molecule. Raman spectroscopy is ideally suited to studying polymorphism as the lack of sample handling minimises the risk of converting the form during measurement as can occur with other techniques. Surprisingly, an early but extensive review by Threlfall [166] of analytical techniques employed to study polymorphism did not have a very large section on Raman spectroscopy. Much of what was reported was with FT Raman. Even so the technique has shown differences with polymorphs of several compounds [167–170] including the B and C forms of naphthazarin [171] and the morphological composition of cimetidine [172]. In addition to that of the active agent, the differing crystallinity of excipients such as glucose has also been recorded [173]. The strict confidentiality surrounding pharmaceutical drug structures is the composition of tablets and this means that much of this work has not been reported in the open literature. For example, a recent application note [174] on polymorphism by dispersive Raman refers only to forms A and B without identifying the drug. One study used low-frequency Raman to characterise the polymorphism of caffeine [160]. A review by Frank [175] contains many examples of the use of Raman spectroscopy in the study of pharmaceuticals.

Yet another example of an unnamed pharmaceutical intermediate is shown in Figure 6.23 recorded by the authors. The FT Raman spectra were recorded of pure crystalline and amorphous forms of the intermediate. A suspect batch of the amorphous form was examined without opening the sampling vial. The spectrum shows evidence of crystallinity at ~1500 cm^{-1}.

More recently the stabilising effect of excipients on the crystallinity of a drug has been studied using model compounds ketoprofen, danazol, griseofluvin and probucol [176].

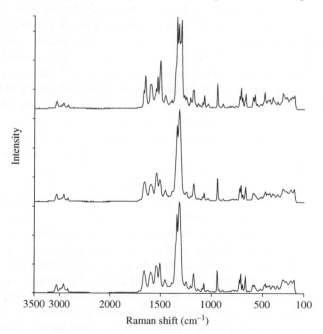

Figure 6.23. Crystallinity in pharmaceutical intermediate: crystalline form (top); amorphous (middle); mixture of crystalline and amorphous forms (bottom).

However, some examples of the use of Raman spectroscopy with pharmaceuticals for which the structure is known have been published. For example, the use of Raman scattering for high-throughput screening of carbamazepine (**6**) has been reported by the Novartis company [177].

(6)
Carbamazepine

This compound has been examined to ascertain the sensitivity of transmission Raman spectroscopy for polymorph detection [178].

6.9   FORENSIC APPLICATIONS

Modern developments in equipment for the detection of Raman scattering make the method very useful in the area of forensic science [179–181]. Advantages include the noninvasive, noncontact nature of the method, the molecularly specific nature of Raman scattering and the speed of analysis. The range of applications is very wide, ranging from detecting drugs and explosives, etc. in the field, either as bulk samples or as traces left at a crime scene, to examination in a clean laboratory of bulk or trace samples such as bags of sugar which may contain cocaine or of single fibres adhering to clothing or other material. Mostly NIR and visible excitation are used and SORS is valuable. The instruments range from rugged handheld devices destined for the back of a police, fire service, military or customs vehicle to sensitive microscopes. Flexible probes can be helpful to reach inaccessible samples and if military applications are included, standoff detection is also useful. The way the results are to be used also varies widely. For example, in the field, detecting drugs and explosives requires an immediate and positive result, as is the case with the use of SORS in an airport to screen for harmful substances through containers. In the fire service, immediate identification of flammable or explosive substances is vital and later detection of substances in residues can be helpful. However, if evidence is to be collected for court use, very careful protection of the scene of crime and the actual evidence (garment, weapon, etc.) is paramount, and this choice affects the instrument and its use significantly.

Many different types of samples are studied. Raman signatures from blood or other biological fluids can give specific signatures which with modern data-processing methods can discriminate between subjects and gunshot residues can determine the nature of the cartridge used. A microfluidics device to detect drugs in saliva has been reported [182]. Analysis of dyes in fibres, paper and inks can establish provenance or demonstrate forgery or simply match a fibre to a specific garment. Microscopes can find very small samples such as a small particle of explosive caught in a fingerprint [183]. Usually fluorescence is used to add security to labels for expensive goods but there is no technical reason why nonfluorescent dye mixtures should not be used for detection and coding with resonance Raman scattering or SERS. In addition to solid and liquid samples, gas-phase samples can be of importance, for example, to collect the vapour from drug or explosive manufacture. In this case some form of collection such as gas-phase chromatography or a cold finger may be required to concentrate the sample although LIDAR, dealt with in Chapter 7, could be used.

SERS involves contact between the sample and the enhancing substrate which may not be permitted in some circumstances. However, with microscopes the area to be covered can be microscopically small and the improved sensitivity and selectivity of SERS can be very useful. It is important that the substrate is effectively applied. For example, colloid is often used. It can be pipetted accurately onto a very small area of a surface aiding discrimination of fibres and inks. If this is the approach, it is important the colloid is applied correctly. It must be concentrated enough to create a solid layer with hot spots but which must be sufficiently thin for the excitation from above to create a plasmon which can penetrate to the sample surface (see the polymer example in Chapter 5). This can greatly increase the sensitivity of Raman scattering making it ideal to study even very small amounts of inks or dyes to identify mixtures for comparison or to identify trace quantities of drugs and explosives. Figure 6.24 shows the spectrum obtained from lipstick smears on glass and cotton where the chromophore gives strong SERRS.

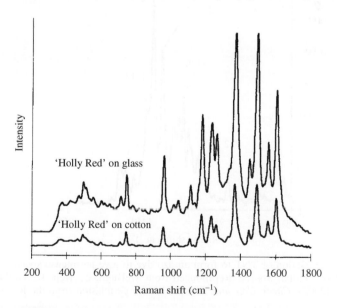

Figure 6.24. SERRS of smears of lipstick on glass and cotton give sharp spectra. Source: Reproduced from Rodger, C., Rutherford, V., Broughton, D., et al. (1998). *Analyst* **123**: 1823 [94] with permission from the Royal Society of Chemistry.

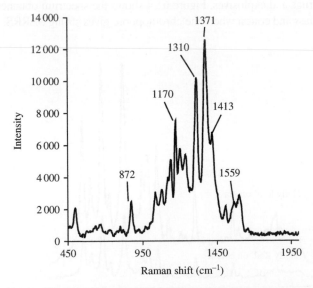

Figure 6.25. Reaction of TNT to form an azo dye to give an intense spectrum. Spectrum shown is at 10^{-9} molar. Source: Reproduced from McHugh, C., Keir, R., Graham, D., and Smith, W.E. (2002). *Chem. Commun.* **580** [184] with permission from the Royal Society of Chemistry.

In some cases, very small quantities such as drugs from a cleaned surface may be required and swabbing methods can be important as can chromatography. SERS molecular specificity and sensitivity can help here. Sometimes SERRS can be helpful. Figure 6.25 shows SERRS of an azo product formed with TNT to give a chromophore and the resulting strong spectrum. This can be done using a flow method and making the colloid *in situ*, thus enabling repeated SERRS measurements of the TNT derivative in real time to give high sensitivity.

6.10 PROCESS ANALYSIS AND REACTION FOLLOWING

6.10.1 Introduction

Whilst many applications mentioned in this chapter so far have used static or *in situ* measurements, a growing area of interest for Raman spectroscopy is reaction following. In principle, the technique is ideal, being noninvasive, able to detect from within glass vessels and aqueous media, and able to carry out monitoring at long distances. Published reports of industrial applications have until recently been relatively sparse. Many have covered laboratory trials or proof of principle. This has been due to several reasons. Fluorescence makes the technique very application-specific. If an instrument is developed to monitor a specific reaction, it does not follow that it can be easily transferred to other applications. The instrumentation was until recently very large, expensive, required specific environmental conditions, i.e. dark rooms, laser interlocked doors. With the introduction of modern variable filter instruments and smaller spectrometers, the instrumentation has become much more user friendly and adaptable. Flexibility of application has increased with laser sources at 785 and 1064 nm and in the UV, thus reducing fluorescence for reaction following. The portability and increased simplicity of the instruments has also opened up the possibility of on-plant monitoring. The latter is somewhat under-reported. This tends to be due to companies wishing to maintain the commercial edge gained from greater efficiencies achieved by tighter plant control. The number of reported examples of reaction following and plant monitoring is increasing. Two excellent reviews have been published [185, 186] on the parameters to be aware of and potential pitfalls in introducing Raman spectroscopy in an industrial plant. A few typical applications are given.

6.10.2 Electronics and Semiconductors

Probably one of the biggest QC applications [187, 188] for Raman spectroscopy is monitoring of the protective diamond-like films (DLF) for computer hard disks. Information on hydrogen content, sp^2/sp^3 ratios and long-range ordering is available from the Raman spectra. Instruments solely dedicated to these measurements can automatically predict the tribological qualities of the films [186, 189]. QC techniques have been established for the physical and chemical characterisation of semiconductors, which can include crystal size and form, dopant levels, stress and strain. This is a major topic in itself for which there are several reviews [190–192] and which is also discussed in Section 6.6. Another feature of process monitoring in this industry is in the deposition and/or growth of thin films. Raman spectroscopy is particularly applicable when the process occurs under vacuum or at elevated temperatures. This could be potentially important for monitoring novel heterostructures [193]. The growth of InSb on Sb (111) has been studied [194] with thicknesses of 0–40 nm. The growth of ZnSe on GaAs at 300 °C followed by capping with Se has been reported [195]. This was followed by crystallisation studies of the Se layer. Other studies [196] have included the nitridation of ZnSe on the GaAs and the growth of CdS on InP (100). This industry probably makes the largest use of process and QC applications of Raman spectroscopy. It is an application which is extremely important to modern life and technology and yet does not greatly feature in conventional literature on analytical chemistry.

6.10.3   PCl$_3$ Production Monitoring

If a plant analyst was asked which reactions would cause the largest problems in monitoring, then elemental phosphorous combined with chlorine in a boiling liquid would feature high on the list. Yet this, the production of PCl$_3$, is one of the best-known examples of Raman online monitoring [196] developed by Freeman et al. The levels of chlorine have to be maintained to prevent the production of PCl$_5$. Using a sampling loop, an FT Raman spectrometer and fibre bundles with laser powers of 2 W, and with 140 scans and 16 cm^{-1} resolution, detection levels of <1% for P$_4$ and Cl$_2$ were attained. Side products of POCl$_3$ and SbCl$_3$ were also monitored but these degraded under the high laser power. Gervasio and Pelletier [197] refined the measurement with a CCD-dispersive system, a 785 nm laser and a direct insertion probe.

6.10.4   Anatase and Rutile Forms of Titanium Dioxide

One of the earliest publications on plant control is the monitoring of the physical form of titanium dioxide, a very bright and white commonly used pigment. The pigment exists in different physical forms, the major ones being anatase and rutile. The rutile form, having a higher opacity, is more commonly used. Both forms have very distinctive bands in the Raman spectrum (see Figure 6.3). The anatase form has bands at 640, 515, 395 and 145 cm^{-1} whilst the rutile bands appear at 610 and 450 cm^{-1}. These bands have been used to quantitatively measure 1% of anatase in rutile with enough accuracy for semiautomated plant control [198]. The greatest difficulty with this measurement was the dusty environment inside the production plant which resulted in major engineering problems. These problems and their solution have been colourfully described at length by Everall et al. [185].

6.10.5 Polymers and Emulsions

As described in Section 6.4, one of the simplest reactions to follow by Raman spectroscopy is the loss of the >C=C< bond. This gives very clear and strong bands in simple monomers such as acrylates, vinyl acetates and styrene. There are numerous literature references covering a range of applications from the simple to the complex. Styrene (S) and methylmethacrylate (MMA) have been studied in a reaction cell [199], homopolymerisations of MMA and butyl acrylate (BuA) have been monitored by FT Raman in laboratory reactors [200] and an analysis of a complex quaternary polymer (S/BuA/MMA/cross-linker) has been carried out [201] to show the consumption of monomers followed by the composition of the resultant copolymer. It would be easy to think that the >C=C< bond band could be monitored quantitatively directly and this can indeed be the case [199, 200]. However, changes in laser intensity, spectrometer response and inhomogeneity can lead to the band having to be normalised. This is usually carried out by choosing a band not affected by the reaction being followed. Unfortunately, this simplistic approach cannot often be used.

Several workers [199, 202] have reported intensity changes in bands, used for reference purposes, to distinguish between the monomer and polymer states. This has been discussed in depth by Everall [203]. Whilst these systems can be monitored quantitatively, similar reactions taking place in emulsions require even greater care. The monomer can exist as droplets, dissolved in water, as micelles or in the polymer phase. The band intensity and wavenumber position can be affected by which phase the monomer is in. Temperature and pressure changes can affect both the spectrum and the phase in which the monomer resides. Equally importantly, the detection of the drops can be affected by the size relative to the wavelength of the laser exciting line (see particle size effects, Chapter 2). These factors need to be taken into consideration but should not prevent analysis taking place, as a published application [204] monitoring latex emulsion polymerisation has shown. Vinyl and acrylate monomers have been largely described here but other systems have also been reported such as cyanate esters [205], epoxies [206], melamine-formaldehydes [207], polyimides [208] and polyurethanes [209]. All the applications described so far have taken place largely in bulk in reaction vessels. Raman spectroscopy has developed in these applications through the use of direct probes and/or coupling with fibre optics which in the case of visible laser sources can be monitored at distances of up to several metres. Another area of application employing *in situ* analysis by Raman spectroscopy is in extruders, fibres and films. In these cases the physical properties of the polymer can be studied as well as the chemical composition. Early work by Hendra [210], taking advantage of polarised Raman spectroscopy, analysed polymers in a lab-scale extruder for information on crystallinity, chain conformation and orientation. Modern instrumentation and fibre-optics developments have enabled similar measurements to be carried out *in situ* on pilot plant and full-scale production facilities. Recently, Chase has also taken advantage of polarised Raman spectroscopy to study fibres at varying points of the drawing process and has carried out extensive studies of fibres on spinning rigs [211, 212].

Similar measurements would appear to be applicable to polymer film production. The measurements are quite complex as the morphological properties can develop in three dimensions. Farquharson and Simpson [213] demonstrated the feasibility with a dispersive spectrometer and a 5 m fibre bundle in a first-reported Raman on-line analysis of polymer film production. Since then, Everall has carried out extensive work on the composition of polyester film on a moving production line using imaging fibre probes coupled through 100 m of fibre [214, 215]. One of the features discovered in this work was the difficulty in handling fluorescence in a moving film compared to static measurements. When a polymer film is static in the beam, low levels of fluorescence can be burnt out (photobleached). With a moving film the sample is rapidly refreshed maintaining the level of fluorescence. Moving from visible excitation towards 785 nm can reduce these effects but Everall found that addition of reclaimed polymer to the stream was a cause of fluorescence [203]. Monitoring polymer film of various types is an application of plant control where Raman spectroscopy would have been expected to have wide use. The lack of literature could be due to commercial sensitivity.

6.10.6 Pharmaceutical Industry

The pharmaceutical industry has many potential applications as described in Section 6.7. The minimal sample handling and the ability to 'see' into polymer containers would be expected to lead to numerous QC applications. Instrument manufacturers claimed expanding sales due to interest in polymorphism alone. Yet little literature, apart from manufacturers' application notes, is available for on-plant applications. Again, it is probably commercial sensitivity which is the cause. A review of biopharmaceuticals manufacturing has been published [216] and a pharmaceutical application has been recently published [217]. Interested readers should also consult the later references in Section 6.7.

6.10.7 Solid-Phase Organic Synthesis/Combinatorial Chemistry

As mentioned in the previous section, one of the growing areas of interest for Raman spectroscopy is solid-phase organic chemistry (SPOC) or combinatorial chemistry. Rather than being carried out in solution, reactions take place in/on solid supports. These supports are usually beads of polyacrylate or polystyrene with reactive end groups such as hydroxyls, chlorine or glycol groups [218]. The beads are porous and have a cellular structure. The beads swell in the solvent and can become gelatinous. The reactants diffuse in and out of the beads with the reactions taking place at the active sites. Analysis of the beads is dependent on the information required (Figure 6.26). Often the phrase 'on bead analysis' is used which leads to the erroneous conclusion that surface techniques apply. However, Raman spectroscopy can penetrate into the bead, enabling effective signals to be obtained. Approximately 90% of the reaction takes place in the bead. For research

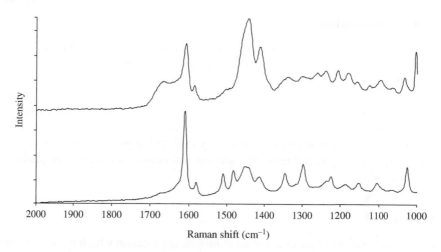

Figure 6.26.   Spectrum of acrylate bead (bottom) and with peptide (top).

purposes single beads are often monitored but for more routine analysis batches
of beads are usually studied. A recent comparison [219] of FTIR and FT Raman
methods highlighted the strengths and weaknesses of both approaches. The advantages of Raman spectroscopy is once again the lack of sample preparation, weak
absorptions from the solvents and the ability to carry out *in situ* studies. The beads
themselves can be studied during preparation as reactive beads, or reactions taking
place in/on the beads can be studied.

$$CH_2 - O - C - Cl$$

(7)
Fmoc-Cl

   Peptide reactions can have several sequences of protection and de-protection.
The 9-fluorenyl-methoxy-carbonyl (Fmoc) (**7**) strategy is an early developed system
[220, 221].

   The stages have been monitored directly on the beads by investigating changes
in the secondary structure [222]. The amide I and III bands were studied to gain
information on the secondary structure of the growing peptide chain. Comparative
studies of *in situ* reactions on beads have been carried out by FTIR and FT Raman
spectroscopy [223–225]. More recently, the use of dispersive Raman spectrometers
has also been investigated for studies with flow through cells [226].

6.10.8   Fermentations

Biotechnology and bioreactors are a fast growing area of technology. The advantage of Raman spectroscopy is the ability to study aqueous systems, though fluorescence and particulates are still potential problems. Little was reported prior to the introduction of lasers in the red end of the spectrum. Shaw et al. [227] have analysed glucose fermentation with a fibre-optic coupled 785 nm laser. They extracted and filtered the liquid to remove yeast cells, which, though not a direct measurement, demonstrated the potential of the technique. By employing PLS techniques and other modelling techniques, including neural networks, glucose and ethanol contents were predicted with errors of ~4%.

6.10.9   Gases

Raman spectroscopy of gases is not an industrial application which readily springs to mind yet, as stated in Chapter 2, some of the earliest Raman spectroscopy was carried out on gases and vapours in inclusions in minerals and rock. The low sensitivity of Raman spectroscopy to gases, due to the low scattering cross-section and few molecules in a given volume, would appear to reduce the applicability. However, these drawbacks have been overcome by using special cells, instruments or remote sensing techniques such as Raman LIDAR [228]. A simple case which demonstrates the sensitivity of Raman spectroscopy to interference is the interference that can occur in weak Raman spectra from fluorescent room lights. Here, sharp emission bands are recorded which can be mistaken for the sample and would certainly cause problems with multivariate analysis routines. The pattern varies with the laser exciting line used. Relatively weak spectra can be enhanced by using multipass cells [189] and placing these inside the laser cavity. Quantitative measurements have been made of CO, CO_2, H_2, H_2O, N_2, N_2O, NH_3 and hydrocarbons using a simple spectrometer constructed from simple components and an intracavity gas cell. The Raman analyser, known as a 'Regap' analyser, has been described by de Groot and Rich [229] for measuring and controlling atmospheres in a steel treatment furnace. Gas analysers have also been developed for the analysis of natural gas [230, 231]. Large-scale QC methods have been devised [232], but methods using Raman microscopes are also used. The applications include accelerator devices for vehicle air bags and degradation of pharmaceuticals inside package products [233]. In each of these applications it is the *in situ* aspects of Raman spectroscopy which overcome the other apparent limitations.

6.10.10   Catalysts

Analysing the surface of catalytic substrates which may be in a glass container, and possibly under water or other solvent is difficult and Raman spectroscopy has obvious advantages in that it is a standoff technique which can give molecularly specific information quickly even under water and the surface can be imaged easily. Not surprisingly, this field generates many papers. Even in 2000, at an American Chemical Society (ACS) Conference, statistics were produced which showed that Raman-based catalysts studies were being published at a rate in excess of 300 a year. Clearly we cannot cover a field of that size adequately in this book but it is well covered in many other publications including Refs. [234–237] which exemplify the main approaches. The simplest advantages have already been stated but Raman spectroscopy can also study systems inside vessels over a range of temperatures and pressures and if both Stokes and anti-Stokes spectra are measured an estimate of temperature at the surface can be obtained. As many catalyst studies are carried out at several hundred degrees centigrade, this is a major factor. Typical studies are in the automotive industry where catalytic converters for vehicle exhausts operate most efficiently at higher temperatures [238]. The crystal structure of metal oxides and metal coordination chemistry is a wide field of study. Metals commonly encountered are platinum, palladium, ruthenium, titanium, uranium, vanadium and zirconium. Alumina- and silica-based catalysts are of continuing wide interest. Many of these materials contain a chromophore which opens up the field to resonance Raman studies. This then has the added advantage of increased selective sensitivity. One of the problems regularly encountered is fluorescence. UV laser sources help overcome fluorescence and increase sensitivity [239] but infrared sources are more readily available and also overcome the fluorescence problem, often with less sample damage. Complete reaction following is possible, but catalytic partial oxidation is also important in industry for production of materials such as alcohols. A typical case is the production of methanol from methane [240].

One of the largest areas of study, of course, is in electrochemical reactions, particularly studies on electrode surfaces. In addition to the normal Raman and resonance Raman advantages in this area, SERS is of great importance. Indeed, as previously shown, the SERS effect was first observed on an electrode surface and often as SERRS is a very powerful tool in this area. For example, when adsorbed on an electrode to control potential it is possible to study the redox cycle of enzymes such as P450 which have a resonant (porphyrin) chromophore at the active site [241].

The increasing use of SERS has led to two new types of catalysts which use the active surface, plasmonic catalysts and enzyme mimetic catalysts or nanozymes. Plasmonic catalysts use the SERS-active surface which when the plasmon is excited, causes a catalytic reaction to occur. The concept is that when a charge transfer state is formed, as well as causing Raman scattering, electron transfer to an excited state of the adsorbate can lead to the electron remaining in a high-energy state of the adsorbate long enough to transfer energy to a second molecule, causing reaction between them. There is plenty of evidence that light-induced reactions occur at the surface such as the chemical transformation of *para*-aminothiophenol to 4,4′-dimercaptoazobenzene [242] which is easily tracked by SERS.

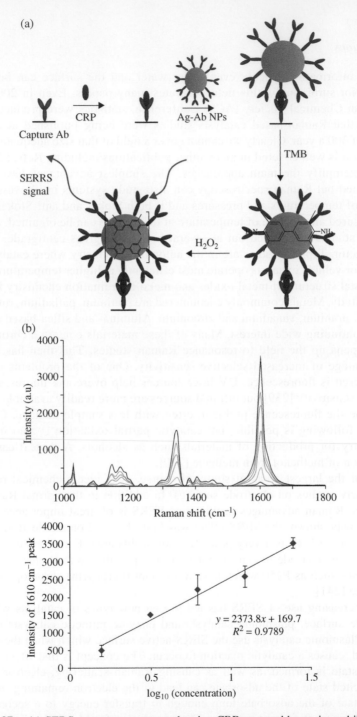

(a)

CRP

Capture Ab

Ag-Ab NPs

TMB

SERRS
signal

H₂O₂

(b)

Figure 6.27. (a) SERS nanozyme concept showing CRP captured by an immobilised antibody enabling the capture of an antibody-coated silver particle. TMB is then added. The silver particle provides the peroxide to react with TMB to give the coloured product. (b) Assay result showing spectra and concentration dependence. Source: Reproduced from Sloan-Dennison, S., Laing, S., Shand, N.C., et al. (2017). *Analyst* **142**: 2484 [245] with permission from the Royal Society of Chemistry.

Frequency-dependent data and high vacuum data make a good case for plasmon involvement [243, 244]. However, most papers ignore any other mechanism. As pointed out in Chapter 5, surface layers such as oxides which can form on silver and copper are photosensitive, so the actual surface reactions could be quite complex and the generation of electrons at a surface with radiation is well known and does not in itself ensure plasmon involvement.

Nanozymes are systems which use the properties of nanoparticles to create artificial enzymes. A standard technique for sensitive detection in antibody assays is to use an enzyme such as horse radish peroxidase to generate hydrogen peroxide *in situ*. This then reacts with a reagent such as TMB to generate a coloured product for detection. In a nanozyme, silver nanoparticles can be used to generate the peroxide making SERS/SERRS an attractive technique for sensitive detection [245]. Figure 6.27 illustrates the development of a nanozyme for SERS detection of C reactive protein (CRP) often analysed in current clinical laboratory practice. A capture antibody is attached to the surface to capture the CRP. A silver nanoparticle coated in a second antibody is then added and it attaches to the CRP. Addition of TMB to the system causes the TMB to react with peroxide from the silver surface to form a coloured product which adsorbed on the nanoparticle gives a strong SERRS signal, high sensitivity and a quantitative response.

6.10.11 Nuclear Industry

Perhaps one of the least likely areas for Raman spectroscopy is in the nuclear processing plants. However, given in noninvasive and remote sensing attributes then it could be thought of as an ideal analytical approach. Applications have been reported of Raman spectroscopic examination of thorium dioxide–uranium dioxide (ThO_2–UO_2) fuel materials and high-level wastes from spent nuclear fuel reprocessing [246, 247].

6.11 SUMMARY

This chapter whilst giving only a flavour of the vast range of applications in which Raman spectroscopy has been utilised should lead the reader to realise the specific advantages of the method. This chapter, together with the previous chapters, shows that, care in matching the instrument and accessories to a specific application can create a very powerful tool which can be of use to both the expert spectroscopist and the general analyst. The next chapter leads into where the technique can yield even more information, albeit requiring, in some cases but not all, expensive, specialist equipment.

REFERENCES

1. Sidorov, N.V., Palatnikov, M.N., Yanichev, A.A. et al. (2016). *J. Appl. Spectrosc.* **83** (5): 750.
2. Li, J.-J., Li, R.-X., Dong, H. et al. (2017). *J. Appl. Spectrosc.* **84** (2): 237–241.
3. Parker, S.F., Refson, K., Bewley, R.I., and Dent, G. (2011). *J. Chem. Phys.* **134**: 084503.
4. Prasetyo, A., Mihailova, B., Suendo, V. et al. (2017)). *J. Raman Spectrosc.* **48** (2): 292–297.
5. Yoshikawa, M. and Nagai, N. (2001). *Handbook of Vibrational Spectroscopy*, vol. 4 (ed. J. Chalmers and P. Griffiths), 2593–2600. New York: Wiley.
6. Dresselhaus, M.S., Dresselhaus, G., Pimenta, M.A., and Eklund, P.C. (1999). *Analytical Applications of Raman Spectroscopy* (ed. M.J. Pelletier), 367–434. Oxford: Blackwell Science.
7. Hendra, P.J. (1996). *Modern Techniques in Raman Spectroscopy* (ed. J.J. Laserna), 94. New York: Wiley.
8. Dhamelincourt, P., Wallart, F., LeClerq, M. et al. (1979). *Anal. Chem.* **51** (4l4A).
9. Yang, Y., Zheng, H., Sun, Q. et al. (2013). *Appl. Spectrosc.* **67** (7): 808–12,.
10. Etz, E.S., Rosasco, G.J., and Cunningham, W.C. (1977). *Environmental Analysis* (ed. G.W. Ewing), 295. New York: Academic Press.
11. Beny, C., Prevosteau, J.M., and Delhaye, M. (1980). *L'actualité Chim.* **April**: 49.
12. Rosasco, C.J. (1978). *Proceedings of the 6th International Conference on Raman Spectroscopy*. London: Heyden.
13. Martoja, M., Tue, V.T., and Elkaim, B. (1980). *J. Exp. Mar. Biol. Ecol.* **43**: 251.
14. G. Dent. *Internet J. Vib. Spectrosc.* http://www.irdg.org/ijvs/reference-spectra/ (accessed 9 October 2018).
15. Wang, A., Han, J., and Guo, L. (1994). *Appl. Spectrosc.* **48**: 8.
16. Nyquist, R.A., Putzig, C.L., and Leugers, M.A. (1997). *IR and Raman Spectral Atlas of Inorganic Compounds and Organic Salts*. Academic Press.
17. Varetti, E.L. and Baran, E.J. (1994). *Appl. Spectrosc.* **48**: 1028.
18. Edwards, D.H.M. and Schnubel, H.J. (1977). *Rev. Gemnol.* **52**: 11.
19. Kawakami, Y., Yamamoto, J., and Kagi, H. (2003). *Appl. Spectrosc* **57**: 1333–1339.
20. Li, L., Du, Z., Zhang, X. et al. (2017). *Appl. Spectrosc.* **72**: 48–59.
21. Popp, J., Tarcea, N., Kiefer, W., et al. (2001). *Proceedings of the First European Workshop on Exo-/Astr-Biology ESA SP-496*. ESRIN, Frascati, Italy.
22. Clegg, S.M., Wiens, R., Misra, A.K. et al. (2014). *Appl. Spectrosc.* **68** (9): 925–936.
23. Angel, S.M., Gomer, N.R., Sharma, S.K., and McKay, C. (2012). *Appl. Spectrosc.* **66** (2): 137–150.
24. R. Frost, T. Kloprogge and J. Schmidt 1999. *Internet J. Vib. Spectrosc.* www.irdg.org/ijvs, 3, 4, 1 (accessed 9 October 2018).
25. Blacksberg, J., Alerstam, E., Maruyama, Y. et al. *Appl. Opt.* **55**: 739.
26. Wei, J., Wang, A., Lambert, J. et al. (2015). *J. Raman Spectrosc.* **46**: 810.
27. Sakurai, T., Ohno, H., Motoyama, H., and Uchida, T. (2017). *J. Raman Spectrosc.* **48** (3): 448–452.
28. https://doi.org/10.1175/AMSMONOGRAPHS-D-15-0026.1
29. Hofer, J., Althausen, D., Abdullaev, S.F. et al. (2017). *Atmos. Chem. Phys.* **17**: 14559–14577.
30. Shintaro, P., Tianxi, Y., and Lili, H. (2016). *Trends Anal. Chem.* **85**: 73–82.
31. Frere, L., Paul-Pont, I., Moreau, J. et al. (2016). *Mar. Pollut. Bull.* **113**: 461.
32. Docherty, J., Mabbott, S., Smith, E. et al. (2015). *Analyst* **140**: 6538–6543.
33. Docherty, J., Mabbott, S., Smith, E. et al. (2016). *Analyst* **141**: 5857.

34. Lombardi, D.R., Wang, C., Sun, B. et al. (1994). *Appl. Spectrosc.* **48**: 875–883.
35. Williams, K. (2000). *Spectroscopy Innovations*, vol. 6. Renishaw Ltd.
36. Fisher, D. and Spits, R.A. (2000). *Gems and Gemology* **Spring**: 42.
37. Phan, D.T.M., Haeger, T., and Hofmeister, W. (2017). *J. Raman Spectrosc.* **48** (3): 453–457.
38. H.F. Shurvell, L. Rintoul and P.M. Fredericks (2001). *Internet J. Vib. Spectrosc.* **5**, 5, 2. www.irdg.org/ijvs (accessed 9 October 2018).
39. Peipetis, A., Vlattas, C., and Galiotis, C. (1996). *J. Raman Spectrosc.* **27**: 519.
40. Madariaga, J.M., Maguregui, M., Castro, K. et al. (2016). *Appl. Spectrosc.* **70** (1): 137–146.
41. Cesaratto, A., Centeno, S.A., Lombardi, J.R. et al. (2017). *J. Raman Spectrosc.* **48** (4): 601–609.
42. Zhao, H.X. and Li, Q.H. (2017). *J. Raman Spectrosc.* **48** (8): 1103–1110.
43. Edwards, H.G.M., Falk, M.J., Sibley, M.G. et al. (1998). *Spectrochim. Acta A* **54**: 903.
44. Clark, R.J.H. (2001). *Handbook of Vibrational Spectroscopy*, vol. 4 (ed. J. Chalmers and P. Griffiths), 2977. New York: Wiley.
45. Perez, F.R., Edwards, H.G.M., Rivas, A., and Drummond, L. (1999). *J. Raman Spectrosc.* **30**: 301.
46. Cesaratto, A., Nevin, A., Valentini, G. et al. (2013). *Appl. Spectrosc.* **67** (11): 1234–1241.
47. Gonzalez-Vidal, J.J., Perez-Pueyo, R., Soneira, M.J., and Ruiz-Moreno, S. (2015). *Appl. Spectrosc.* **69** (3): 314–322.
48. Conti, C., Colombo, C., Realini, M. et al. (2014). *Appl. Spectrosc.* **68** (6): 686–691.
49. Zuo, J., Xu, C., Wang, C., and Yushi, Z. (1999). *J. Raman Spectrosc.* **30**: 1053.
50. Clark, R.J.H. (1995). *J. Mol. Struct.* **347**: 417–428.
51. Yu, J. and Butler, I.S. (2015). *Appl. Spectrosc. Rev.* **50** (2): 152–157.
52. Dele, M.L., Dhamelincourt, P., Poroit, J.P., and Schnubel, H.J. (1986). *J. Mol. Struct.* **143**: 135.
53. Barone, G., Mazzoleni, P., Raneri, S. et al. (2016). *Appl. Spectrosc.* **70** (9): 1420–1431.
54. Clark, R.J.H., Curri, M.L., and Largana, C. (1997). *Spectrochim. Acta* **53A**: 597.
55. McCann, L.I., Trentleman, K., Possley, T., and Golding, B. (1999). *J. Raman Spectrosc.* **30**: 121.
56. Edwards, H.G.M., Farewell, D.W., and Quye, A. (1997). *J. Raman Spectrosc.* **28**: 243.
57. Edwards, H.G.M., Hunt, D.E., and Sibley, M.G. (1998). *Spectrochim. Acta* **54**: 745.
58. Carter, E.A. and Edwards, H.G.M. (2001). *Infrared and Raman Spectroscopy of Biological Materials* (ed. H.-U. Gramlich and B. Yan). New York: Marcel Dekker.
59. Chalmers, J. and Griffiths, P. (eds.) (2001). *Handbook of Vibrational Spectroscopy*, vol. 4. New York: Wiley.
60. Hendra, P.J. and Agbenyega, J.K. (eds.) (1993). *The Raman Spectra of Polymers*. Wiley.
61. Schrader, B. (1989). *Raman/Infrared Atlas of Organic Compounds*, 2e. Weinheim: Wiley-VCH.
62. Garton, A., Batchelder, D.N., and Cheng, C. (1993). *Appl. Spectrosc.* **47** (7): 922.
63. Chalmers, J.M. and Everall, N.J. (1993). *Polymer Characterisation* (ed. B.J. Hunt and M.I. James). Glasgow: Blackie Academic.
64. Cornell, S.W. and Koenig, J.L. (1969). *Macromolecules* **2**: 540.
65. Frankland, J.A., Edwards, H.G.M., Johnson, A.F. et al. (1991). *Spectrochim. Acta* **47A**: 1511.
66. Jackson, K.D.O., Loadman, M.J.R., Jones, C.H., and Ellis, G. (1990). *Spectrochim. Acta* **46A**: 217.
67. Tashiro, K., Ueno, Y., Yoshioka, A. et al. (1999). *Macromol. Symp.* **114**: 33.

68. Tashiro, K., Sasaki, S., Ueno, Y. et al. (2000). *Korea Polym. J.* **8**: 103.
69. Everall, N.J., Chalmers, J.M., Kidder, L.H. et al. (2000). *Polym. Mater. Sci. Eng.* **82**: 398–399.
70. Sue, H.-J., Earls, J.D., Hefner, R.E. Jr. et al. (1998). *Polymer* **39**: 4707.
71. Walton, J.R. and Williams, K.P.J. (1991). *Vib. Spectrosc.* **1**: 239.
72. Chike, K.E., Myrick, M.L., Lyon, R.E., and Angel, S.M. (1993). *Appl. Spectrosc.* **47**: 1631.
73. Kawagoe, M., Takeshima, M., Nomiya, M. et al. (1999). *Polymer* **40**: 1373.
74. Kawagoe, M., Hashimoto, S., Nomiya, M. et al. (1999). *J. Raman Spectrosc.* **30**: 913.
75. Gerrard, D.L. and Maddams, W.F. (1975). *Macromolecules* **8**: 55.
76. Baruya, A., Gerrard, D.L., and Maddams, W.F. (1983). *Macromolecules* **16**: 578.
77. Owen, E.D., Shah, M., Everall, N.J., and Twigg, M.V. (1994). *Macromolecules* **27**: 3436.
78. Schaffer, H.E., Chance, R.R., Sibley, R.J. et al. (1991). *J. Phys. Chem.* **94**: 4161.
79. Chalmers, J.M. and Dent, G. *Industrial Analysis with Vibrational Spectroscopy*, 1997. London: Royal Society of Chemistry.
80. Rodger, C., Smith, W.E., Dent, G., and Edmondson, M. (1996). *J. Chem. Soc. Dalton Trans.* **5**: 791–799.
81. Persaund, I. and Grossman, W.E.L. (1993). *J. Raman Spectrosc.* **24**: 107.
82. Majoube, M. and Henry, M. (1991). *Spectrochim. Acta A* **47**: 1459.
83. Neipp, K., Wang, Y., Desari, R.R., and Field, M.S. (1995). *Appl. Spectrosc.* **49**: 780.
84. Graham, D., Smith, W.E., Lineacre, A.M.T. et al. (1997). *Anal. Chem.* **69**: 4703–4707.
85. Graham, D., Mallinder, B.J., and Smith, W.E. (2000). *Angew. Chem. Int. Ed. Engl.* **6**: 1061–1063.
86. Graham, D., Mallinder, B.J., and Smith, W.E. (2000). *Biopolymers(Biospectroscopy)* **112**: 1103–1105.
87. Bourgeois, D. and Church, S.P. (1990). *Spectrochim. Acta A* **46**: 295.
88. Everall, N. (1993). *Spectrochim. Acta A* **49**: 727–730.
89. McGeorge, G., Harris, R.K., Chippendale, A.M., and Bullock, J.F. (1996). *J. Chem. Soc. Perkin Trans.* **2**: 1733.
90. McGeorge, G., Harris, R.K., Bastanov, A.S. et al. (1998). *J. Chem. Soc. Perkin Trans.* **102**: 3505–3513.
91. White, P.C., Rodger, C., Rutherford, V. et al. (1998). *SPIE* **3578**: 77.
92. White, P.C., Munro, C.H., and Smith, W.E. (1996). *Analyst* **121**: 835.
93. Was-Gubala, J. and Starczak, R. (2015). *Appl. Spectrosc.* **69** (2): 296–303.
94. White, P.C., Rodger, C., Rutherford, V. et al. (1998). *Analyst* **123**: 1823.
95. Drake, J.A.G. (ed.) (1993). *Chemical Technology in Printing Systems*. London: Royal Society of Chemistry.
96. Kivioja, A., Hartus, T., Vuorinen, T. et al. (2013). *Appl. Spectrosc* **67** (6): 661–671.
97. Rodger, C. (1997). The development of SERRS as a quantitative and qualitative analytical technique. PhD dissertation, University of Strathclyde, Glasgow.
98. Rodger, C., Dent, G., Watkinson, J., and Smith, W.E. (2000). *Appl. Spectrosc.* 54.
99. Armstrong, D.R., Clarkson, J., and Smith, W.E. (1995). *J. Phys. Chem.* **99**: 17825.
100. Mullen, K.I., Wang, D.X., Crane, L.G., and Carron, K.T. (1992). *Anal. Chem.* **64**: 930–936.
101. Zollinger, H. (1991). *Colour Chemistry*. Weinheim: VCH.
102. Venkataraman, K. (1977). *The Analytical Chemistry of Synthetic Dyes*. New York: Wiley.
103. Dent, G. and Farrell, F. (1997). *Spectrochim. Acta* **53A** (1): 21.

104. Wood, S., Hollis, J.R., and Kim, J.-S. (2017). *J. Phys. D: Appl. Phys.* **50**: 73001.
105. Tsumura, A., Koezuka, H., and Ando, T. (1986). *Appl. Phys. Lett.* **49**: 1210.
106. Burroughs, J.H., Jones, C.A., and Friend, R.H. (1988). *Nature* **335**: 137.
107. Bao, Z., Rodgers, J.A., and Katz, H.E. (1999). *J. Mater. Chem.* **9**: 1895.
108. Yu, G., Gao, J., Hummelen, J.C. et al. (1995). *Science* **270**: 1789.
109. Burroughs, J.H., Bradley, D.D.C., Brown, A.R. et al. (1990). *Nature* **347**: 539.
110. Friend, R.H., Gymer, R.W., Holmes, A.B. et al. (1999). *Nature* **397**: 121.
111. Skotheim, T.A., Elsenbaummer, R.L., and Reynolds, J.R. (eds.) (1997). *Handbook of Conducting Polymers*. New York: Marcel Dekker.
112. Sariciftci, N.S. (ed.) (1997). *Primary Photoexcitations in Conjugated Polymers: Molecular Exciton versus Semiconductor Band Model*. Singapore: World Scientific.
113. Keiss, H. (ed.) (1992). *Conjugated Conducting Polymers*. Berlin: Springer-Verlag.
114. Shirota, Y. (2000). *J. Mater. Chem.* **10**: 1.
115. Aldrich Online Chemical Catalogue. (1996). www.sigmaaldrich.com/Brands/Aldrich/Polymer_Products/Specialty_Areas.html (accessed 9 October 2018).
116. Becker, H., Spreitzer, H., Kreuder, W. et al. (2000). *Adv. Mater.* **12**: 42.
117. Bérnard, S. and Yu, P. (2000). *Adv. Mater.* **12**: 48.
118. Su, W.P., Schrieffer, J.R., and Heeger, H.J. (1980). *Phys. Rev. B* **22**: 2099.
119. Su, W.P. and Schrieffer, J.R. (1980). *Proc. Natl. Acad. Sci. USA* **77**: 5626.
120. Brédas, J.L., Chance, R.R., and Sibley, R. (1981). *Mol. Cryst. Liq. Cryst.* **77**: 253.
121. Takabayashi, S., Ješko, R., Shinohara, M. et al. (2018). *Surf. Sci.* **668**: 36.
122. Milani, A., Tommasini, M., Russo, V. et al. (2015)). *J. Nanotechnol* **6**: 480.
123. Merlen, A., Buijnstersand, J.G., and Pardanaud, C. (2017). *Coatings* **7**: 153.
124. Jorio, A. and Souza Filho, A.G. (2016). *Ann. Rev. Mater. Res.* **46**: 357.
125. Wu, J.-B., Lin, M.-L., Cong, X. et al. (2018). *Chem. Soc. Rev.* **47**: 1822.
126. Ferrari, A.C. (2007)). *Solid State Commun.* **143**: 47.
127. Bergman, L. and Nemanich Ann, R.J. (1996). *Rev. Mater. Sci.* **26**: 551.
128. Falkovsky, L.A. (2004)). *Phys Uspekhi* **47**: 249.
129. Gobel, A., Ruf, T., Fischer, T.A. et al. (1999). *Phys. Rev. B* **59**: 12612.
130. Yoshikawa, M. and Ngai, N. (2001). *Handbook of Vibrational Spectroscopy*, vol. 4 (ed. J. Chalmers and P. Griffiths), 2604. New York: Wiley.
131. Schaeberle, M.D., Tuschel, D.D., and Treado, P.J. (2001). *Appl. Spectrosc.* **55**: 257–266.
132. Cerdeira, F., Fjeldly, T.A., and Cardona, M. (1973). *Phys. Rev. B* **8**: 4734.
133. Yoshikawa, M., Agawam, K., Morita, N. et al. (1997). *Thin Solid Films* **310**: 167.
134. Kim, J.-H., Seo, S.-H., Yun, S.-M. et al. (1996). *Appl. Phys. Lett.* **68**: 1507.
135. Harima, H. (2002). *J. Phys.: Condens. Matter* **14**: R967.
136. Byrne, H.J., Knief, P., Keating, M.E., and Bonnier, F. (2016). *Chem. Soc. Rev.* **45**: 1865.
137. Dick, S. and Bel, S.E.J. (2017)). *Faraday Discuss* **205**: 517.
138. Zhao, Z., Shen, Y., Hu, F., and Min, W. (2017). *Analyst* **142**: 4018.
139. Freire, P.T.C. (2000). *Proceedings of the International Conference on Raman Spectroscopy* (ed. S.L. Zhang and B.F. Zhu), 440. Wiley.
140. Schuster, K.C., Reese, I., Urlab, E. et al. (2000). *Anal. Chem.* **72**: 5529.
141. Alexander, T.A., Pelligrino, P.M., and Gillespie, J.B. (2003). *Appl. Spectrosc.* **57**: 1340–1345.
142. Wold, J.P., Marquardt, B.J., Dable, B.K. et al. (2004). *Appl. Spectrosc.* **58**: 395–403.
143. Sockalingum, G.D., Lamfarraj, H., Beljebbar, A. et al. (1999). *SPIE* **3608**: 185.

144. Arcangeli, C. and Cannistraro, S. (2000). *Biopolymers* **57**: 179–186.
145. Piot, O., Autran, J.C., and Manfait, M. (2001). *J. Cereal Sci.* **34**: 191–205.
146. Matsi, H. and Pan, S. (2000). *J. Phys. Chem. B* **104**: 8871.
147. Zheng, J., Zhou, Q., Zhou, Y. et al. (2002). *J. Electroanal. Chem.* **530**: 75–81.
148. Neal, S.L. (2018)). *Appl. Spectrosc.* **72**: 102.
149. Smith, R., Wright, K.L., and Ashton, L. (2016)). *Analyst* **141**: 3590.
150. Ando, J., Palonpon, A.F., Sodeoka, M., and Fujita, K. (2016). *Curr. Opin. Chem. Biol.* **33**: 16.
151. Ember, K.J.I., Hoeve, M.A., McAughtrie, S.L. et al. (2017). *Regenerative Med.* **2**: 12.
152. Ellis, D.I. and Goodacre, R. (2006). *Analyst* **131**: 875.
153. Matousek, P. and Stone, N. (2016). *Chem. Soc. Rev.* **45**: 1794.
154. Kong, K., Kendall, C., Stone, N., and Notingher, I. (2015). *Adv. Drug Deliv. Rev.* **89**: 121.
155. Brusatori, G., Auner, T., Noh, L. et al. (2017). *Neurosurg. Clin. N Am.* **28**: 633.
156. Nicolson, F., Jamieson, L.E., Mabbott, S. et al. (2018). *Chem. Sci.* **9**: 3788.
157. Ellis, G., Hendra, P.J., Hodges, C.M. et al. (1989). *Analyst* **114**: 1061–1066.
158. Pelletier, M.J. (2013). *Appl. Spectrosc.* **67** (8): 829–840.
159. Olds, W.J., Sundarajoo, S., Selby, M. et al. (2012). *Appl. Spectrosc.* **66** (5): 530–537.
160. Larkin, P.J., Dabros, M., Sarsfield, B. et al. (2014). *Appl. Spectrosc.* **68** (7): 758–776.
161. Pelletier, M.J., Larkin, P., and Santangelo, M. (2012). *Appl. Spectrosc.* **66** (4): 451–457.
162. Sparen, A., Hartman, M., Fransson, M. et al. (2015). *Appl. Spectrosc.* **69** (5): 580–589.
163. Everall, N., Priestnall, I., Dallin, P. et al. (2010). *Appl. Sectrosc.* **64** (5): 476–484.
164. Chalmers, J. and Griffiths, P. (eds.) (2001). *Handbook of Vibrational Spectroscopy*, vol. 5. New York: John Wiley & Sons, Inc.
165. W.C. McCrone, in: *Physics and Chemistry of the Organic Solid State*, D. Fox, M.M. Labes and A. Weissberger (eds), vol. II, Interscience, New York, 1965, p. 275.
166. Threlfall, T.L. (1995). *Analyst* **120**: 2435.
167. Anwar, J., Tarling, S.E., and Barnes, P. (1989). *J. Pharm. Sci.* **78**: 337.
168. Neville, G.A., Beckstead, H.D., and Shurvell, H.F. (1992). *J. Pharm. Sci.* **81**: 1141.
169. Deeley, C.M., Spragg, R.A., and Threlfall, T.L. (1991). *Spectrochim. Acta* **47**: 1217.
170. Tudor, A.H., Davies, M.C., Melia, C.D. et al. (1991). *Spectrochim. Acta* **47**: 1389.
171. Paul, S., Schutte, C.H.J., and Hendra, P.J. (1990). *Spectrochim. Acta* **46**: 323.
172. Jalsovszky, G., Egyed, O., Holly, S., and Hegedus, B. (1995). *Appl. Spectrosc.* **49** (8): 1142.
173. Hendra, P.J. (1996). *Modern Techniques in Raman Spectroscopy* (ed. J.J. Laserna), 89. Wiley.
174. Application Note, Polymorph Analysis by Dispersive Raman Spectroscopy, Nicolet, AN119 (2001).
175. Frank, C. (1999). *Analytical Applications of Raman Spectroscopy* (ed. M.J. Pelletier), 224–275. Oxford: Blackwell Science.
176. Chen, X., Stoneburner, K., Ladika, M. et al. (2015). *Appl. Spectrosc.* **69** (11): 1271–1280.
177. Hilfiker, R., Berghausen, J., Marcolli, C. et al. (2002). *Eur. Pharm. Rev.* **2**: 37–43.
178. Feng, H., Anderson, C.A., Drennen, J.K. 3rd et al. (2017). *Appl. Spectrosc.* **71** (8): 1856–1867.
179. Smith, W.E., Rodger, C., Dent, G., and White, P.C. (2001). *Handbook of Raman Spectroscopy* (ed. I.R. Lewis and H.G. Edwards). Marcel Dekker.
180. Doty, K.C., Muro, C.K., Bueno, J. et al. (2016). *J. Raman Spectrosc.* **47**: 39.
181. Muehlethaler, C., Leona, M., and Lombardi, J.R. (2016). *Anal. Chem* **88**: 152.
182. Andreou, C., Hoonejani, M.R., Barmi, M.R. et al. (2013). *ACS Nano* **7**: 7157.

183. Cheng, C., Kirkbride, T.E., Bachelder, D.N. et al. (1995). *J. Forensic Sci.* **40**: 31.
184. McHugh, C., Keir, R., Graham, D., and Smith, W.E. (2002). *Chem. Commun* 580.
185. Everall, N.J., Clegg, I.M., and King, P.W.B. (2001). *Handbook of Vibrational Spectroscopy*, vol. 4 (ed. J. Chalmers and P. Griffiths), 2770–2801. Wiley.
186. I.R. Lewis, in: *Handbook of Raman Spectroscopy*, I.R. Lewis and H.G.M. Edwards (eds), Marcel Dekker, New York, 2001, pp. 919–974.
187. Plano, S. and Adar, F. (1987). *Proc. SPIE* **822**: 52.
188. Tsai, H.C. and Bogy, D.B. (1987). *J. Vac. Sci. Technol. A* **5**: 3287.
189. Adar, F., Geiger, R., and Noonan, J. (1997). *Appl. Spectrosc. Rev.* **32**: 45.
190. Pollack, F.H. (1991). *Analytical Raman Spectroscopy* (ed. J.G. Grasselli and B.J. Bulkin), 137–221. New York: Wiley.
191. de Wolf, I. (1999). *Analytical Applications of Raman Spectroscopy* (ed. M.J. Pelletier), 435–472. Oxford: Blackwell Science.
192. Nakashima, S. and Harima, H. (2001). *Handbook of Vibrational Spectroscopy*, vol. 4 (ed. J. Chalmers and P. Griffiths), 2637–2656. Wiley.
193. Wagner, V., Ritcher, W., Geurtus, J. et al. (1996). *J. Raman Spectrosc.* **27**: 265.
194. Wagner, V., Drews, D., Esser, N. et al. (1994). *J. Appl. Phys.* **75**: 7330.
195. Drews, D., Schneider, A., Zahn, D.R.T. et al. (1996). *Appl. Surf. Sci.* **104/105**: 485.
196. a Zahn, D.R.T. (1998). *Appl. Surf. Sci.* **123/124**: 276. b Freeman, J.J., Fisher, D.O., and Gervasio, G.J. (1993). *Appl. Spectrosc.* **47**: 1115.
197. Gervasio, G.J. and Pelletier, M.J. (1997). *At-Process* **3**: 7.
198. Besson, J.P., King, P.W.B., Wilkins, T.A., et al. (1997). Calcination of titanium dioxide. European Patent EP 0 767 222 A2.
199. Gulari, E., McKeigue, K., and Ng, K.Y.S. (1984). *Macromolecules* **17**: 1822.
200. Clarkson, J., Mason, S.M., and Williams, K.P.J. (1991). *Spectrochim. Acta* **47A**: 1345.
201. Everall, N. and King, B. (1999). *Macromolecules* **141**: 103.
202. Wang, C., Vickers, T.J., Schlenoff, J.B., and Mann, C.K. (1992). *Appl. Spectrosc.* **46**: 1729.
203. Everall, N. (1999). *Analytical Applications of Raman Spectroscopy* (ed. M.J. Pelletier), 127–192. Oxford: Blackwell Science.
204. Bauer, C., Anram, B., Agnely, M. et al. (2000). *Appl. Spectrosc.* **54**: 528.
205. Cooper, J.B., Vess, T.M., Campbell, L.A., and Jensen, B.J. (1996). *J. Appl. Polym. Sci.* **62**: 135.
206. Aust, J.F., Booksh, K.S., Stellman, C.M. et al. (1997). *Appl. Spectrosc.* **51**: 247.
207. Scheepers, M.L., Gelan, J.M., Carleer, R.A. et al. (1993). *Vib. Spetrosc.* **6**: 55.
208. Cooper, J.B., Wise, K.L., and Jensen, B.J. (1997). *Anal. Chem.* **69**: 1973.
209. Xu, L., Li, C., and Ng, K.Y.S. (2000). *J. Phys. Chem. A* **104**: 3952.
210. Hendra, P.J., Morris, D.B., Sang, R.D., and Willis, H.A. (1982). *Polymer* **23**: 9.
211. Chase, D.B. (1996). *XVth International Conference on Raman Spectroscopy* (ed. S. Asher and P. Stein), 1072. Pittsburgh: Wiley.
212. Chase, D.B. (1997). *Mikrochim. Acta* **14**: 1.
213. Farquharson, S. and Simpson, S.F. (1992). *Proc. SPIE* **1681**: 276.
214. Everall, N. (1995). *An Introduction to Laser Spectroscopy* (ed. D.L. Andrews and A.A. Demidov). New York: Plenum Press.
215. Everall, N., King, B., and Clegg, I. (2000)). *Chem. Britain* **July**: 40.
216. Buckley, K. and Ryder, A.G. (2017). *Appl. Spectrosc.* **71** (6): 1085–1116.

217. Grymonpre, W., Bostijn, N., Herck, V.S. et al. (2017). *Int. J. Pharm. (Amsterdam, Netherlands)* **531** (1): 235–245.
218. Atherton, E., Clive, D.L., and Sheppard, R.C. (1975). *J. Am. Chem. Soc.* **97**: 6584.
219. Yan, B., Gremlich, H.-U., Moss, S. et al. (1999). *J. Comb. Chem.* **1**: 46–54.
220. Chang, C.-D. and Meisenhofer, J. (1978). *Int. J. Protein Res.* **11**: 246.
221. Atherton, E., Fox, H., Harkiss, D. et al. (1978). *J. Chem. Soc. Chem. Commun.* 537.
222. Ryttersgaard, J., Due Larsen, B., Holm, A. et al. (1997). *Spectrochim. Acta A* **53**: 91–98.
223. Pivonka, D.E., Russell, K., and Gero, T.W. (1996). *Appl. Spectrosc.* **50**: 1471.
224. Pivonka, D.E., Palmer, D.L., and Gero, T.W. (1999). *Appl. Spectrosc.* **53**: 1027.
225. Pivonka, D.E. (2000). *J. Comb. Chem.* **2**: 33–38.
226. Application Note, In-situ Analysis of Combinatorial Beads by Dispersive Raman Spectroscopy, Nicolet, AN-00121 (2001).
227. Shaw, A.D., Kaderbhal, N., Jones, A. et al. (1999). *Appl. Spectrosc.* **53**: 1419.
228. Renaut, D., Pourny, J.C., and Capitini, R. (1980). *Optics Lett.* **5**: 233.
229. de Groot, W. and Rich, R. (1999). *Proc. SPIE* **3535**: 32.
230. Petrov, D.V. and Matrosov, I.I. (2016). *Appl. Spectrosc.* **70** (10): 1770–1776.
231. Buldakov, M.A., Korolev, B.V., Matrosov, I.I. et al. (2013). *J. Appl. Spectrosc.* **80** (1): 124–128.
232. Weber, W.H., Zanini-Fisher, M., and Pelletier, M.J. (1997). *Appl. Spectrosc.* **51**: 123.
233. Gilbert, A.S., Hobbs, K.W., Reeves, A.H., and Hobson, P.P. (1994). *Proc. SPIE* **2248**: 391.
234. Wachs, I.E. (2001). *Handbook of Raman Spectroscopy* (ed. I.R. Lewis and H.G.M. Edwards). New York: Marcel Dekker.
235. Wachs, I.E. (2013). *Dalton Trans.* **42**: 11762.
236. Stavitski, E. and Weckhuysen, B.M. (2010). *Chem. Soc. Rev.* **39**: 4615.
237. Wachs, I.E. and Roberts Chem, C.A. (2010). *Soc. Rev.* **39**: 5002.
238. Uy, D., O'Neill, A.E., Xu, L. et al. (2003). *Appl. Catal. B* **41**: 269–278.
239. La Parola, V., Deganello, G., Tewell, C.R., and Venezia, A.M. (2002). *Appl. Catal. A* **235**: 171–180.
240. Hutchings, G.J., Hargreaves, J.S.J., Joyner, R.W., and Taylor, S.H. (1997). *Studies Surf. Sci. Catal.* **107**: 41–46.
241. Kitahama, Y. and Ozaki, Y. (2016). *Analyst* **141**: 5020.
242. Huang, Y.-F., Zhu, H.-P., Liu, G.-K. et al. (2010). *J. Am. Chem. Soc.* **132**: 9244.
243. Dong, B., Fang, Y., Chen, X. et al. (2011). *Langmuir* **27**: 10677.
244. Zhang, Z., Deckert-Gaudig, T., and Deckert, V. (2015). *Analyst* **140**: 4325.
245. Sloan-Dennison, S., Laing, S., Shand, N.C. et al. (2017). *Analyst* **142** (2484).
246. Rao, R., Bhagat, R.K., Salke, N.P., and Kumar, A. (2014). *Appl. Spectrosc.* **68** (1): 44–48.
247. Stefanovsky, S.V., Myasoedov, B.F., Remizov, M.B., and Belanova, E.A. (2014). *J. Appl. Spectrosc.* **81** (4): 618–623.

Chapter 7

More Advanced Raman Scattering Techniques

7.1 INTRODUCTION

Modern developments in optics, electronics and software are continually improving the performance of Raman spectrometers and expanding the opportunities for their effective use. Today, robust matchbox-size portable spectrometers, some of which run off 1.5 V batteries give good performance. Software developments have also improved areas such as performance, libraries, fluorescence rejection, signal retrieval from a matrix, data analysis and 3D high resolution and imaging. Effective pulsed tunable lasers with modern optics and improved detectors make multi-phonon techniques such as stimulated Raman scattering (SRS), hyper Raman scattering (HRS) and coherent anti-Stokes Raman scattering (CARS) simpler and more reliable thus making it easier to benefit from their advantages. Coupling between Raman spectrometers and other instruments such as atomic force microscopes (AFM) and electron microscopes (EM) is simpler [1]. Techniques such as tip-enhanced Raman scattering (TERS), Raman optical activity (ROA) and UV Raman scattering have developed to give very valuable results uniquely available using these methods. For many Raman spectroscopists most of these methods will not be readily available, but as the technology becomes simpler and more reliable, access is expanding. The purpose of this chapter is to briefly describe some of these techniques to illustrate the advantages of each and by doing so acquaint the reader with the additional scope they can provide to Raman spectroscopists.

Modern Raman Spectroscopy: A Practical Approach, Second Edition. Ewen Smith and Geoffrey Dent.
© 2019 John Wiley & Sons Ltd. Published 2019 by John Wiley & Sons Ltd.

7.2   FLEXIBLE OPTICS

Portable systems are now widely used, for example, in process analysis or to detect drugs and explosives, environmental pollutants and hazardous materials. They have been made rugged for army, fire service and police use. Software is good and includes dedicated libraries so that an instrument can be programmed simply to set off an alarm or to identify a compound. SORS handheld systems (see Chapter 2) are now being deployed successfully to detect substances such as hazardous materials in containers without opening them (Figure 7.1).

Good fibre optics and probe head design mean that Raman systems including portable systems can be used very flexibly, both in separating the spectrometer from the probe in difficult environments and in coupling spectrometers to other equipment (see Chapter 2). It is also possible to use Raman scattering with other optical arrangements. For example, the laser beam instead of being launched into a fibre optic can be wave guided through a narrow tube filled with a solution containing the substance to be detected. The materials used and the dimensions for the construction of these tubes have to be optimised so that the beam is repeatedly reflected inwards to give wave guiding. When set up correctly, the laser beam can travel through many metres of the sample exciting Raman scattering as it passes. The beam exits at the far end of the tube along with the associated Raman scattered radiation. This huge path length can significantly increase the signal to noise ratio of the Raman scattering.

The increased flexibility and robustness has led to the combination of Raman spectroscopy with many other techniques using fibre-optic-coupled Raman probes or modified microscope systems. These include chromatograph techniques such as TLC, HPLC, CE, GPC, FIA and GC. The advantage is that the analyte can be positively identified from the Raman scattering but works best with analytes which have a large scattering cross-section and in conditions with minimal fluorescence.

Figure 7.1. A handheld SORS spectrometer designed to detect hazardous materials. Source: © Agilent Technologies, Inc. 2016, 2018 Reproduced with Permission, Courtesy of Agilent Technologies, Inc.

Raman spectrometers have also been coupled to differential scanning calorimeters (DSC) to follow changes in the spectrum with changing temperature and to electrochemical cells to follow changes with voltage.

Combination of an EM and a Raman spectrometer gives the advantage that Raman scattering can be obtained for very small objects for which the structure is fully defined by the EM. This saves the problem of matching results from a sample taken with two techniques on two very different dimensions. There is an obvious potential for sample degradation with the use of a high vacuum and irradiation from both a beam of electrons and radiation from a visible or near-IR laser. The combination of Raman spectroscopy with an AFM or a scanning tunnelling microscope (STM) is more widely used. Atmospheric control is easier and, in the form of TERS, this method has given remarkable results which are discussed in Section 7.5.

The quality and flexibility of modern optics make it simpler to optically trap and manipulate particles and cells and to measure their Raman spectrum. In its simplest form, trapping is achieved by tightly focusing an intense beam of light onto a particle. In some particles, such as silica particles, the beam passes through and then reemits. This creates forces on the surface at the point of entry and at the bottom, and the direction of these forces holds the particle in the beam thus trapping or 'tweezing' it. The trapped particle can then be manipulated by using external optics to move the beam so that the particle is placed in a suitable position. Raman spectra can be obtained from these particles simply by collecting the scattered light created by interaction with the trapping beam or with a second beam. Trapping can be used, for example, to bring together two particles or cells to study the interactions. Various uses of tweezing have been reviewed [2].

This system can be set up to give SERS/SERRS spectra and there are ways to trap the SERS-active nanoparticles [3]. However, there is a potential problem with larger particles which are easily observed in the microscope because the light tends to travel round the metal surface rather than through it. For example, larger silver-coated silica particles do not normally trap easily but if they are only lightly coated with silver, the beam is transmitted through the particle, traps well and gives SERS. Figure 7.2 shows SERRS from a lightly coated particle treated with an azo dye and trapped in a tweezer system using 532 nm excitation. One remarkable feature of this is that the scattering is so intense that it can be viewed through a microscope with a standard video camera although the signal decays quickly.

The very small spots which can be used to excite and collect Raman scattering, the ability to focus through glass or polymer and the standoff nature of the technique make it very effective as a detection method for lab-on-a-chip devices. In some cases, it will be sufficient for Raman scattering to be used directly but often the additional sensitivity and selectivity of resonance Raman scattering or more often SERS may be required. With SERS, a decision has to be made as to whether to use a substrate either inserted into the device or made in it or to use colloid which can be circulated through the device when required. For a single-use device, a substrate which is incorporated into the chip can be very effective. However, one of the major advantages of

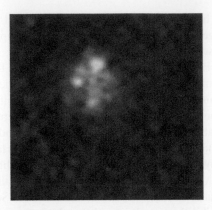

Figure 7.2. SERRS from a single silver-coated silica particle detected with a standard video recorder indicating a short-lived pulse of Raman scattering. The particle was trapped using 1064 nm radiation and the SERRS from an azo dye obtained using 532 nm excitation.

going to the trouble of creating a lab-on-a-chip device is the reproducibility created by standard dimensions and controlled flow rates. Therefore, good reproducibility and shelf life of the substrate are key requirements. For reusable devices, *in situ* substrates are easily poisoned and can be difficult to clean for reuse to give a similar level of performance each time with an electrode being possibly the best approach. Alternatively, colloid can form a very stable thin line which is reproducible due to the control chips provide and which can be imaged to allow continuous measurement. Colloid can be made *in situ*. Figure 7.3 shows the formation of a colloid on chip and the resulting thin linear stream [4]. In most SERS, colloid made with borohydride is seldom used now since it has a short life, but here, the controlled nature of the flow in the chip enables repeat analysis with good reproducibility. There are other advantages which can be built into a chip such as defined path dimensions, mixing chambers and different flow rates in different channels. In addition, samples can be trapped for analysis in various ways. One simple way is to add a set of pillars to the device, as shown in Figure 7.3, which will stop beads with adsorbed analyte and allow the rest of the sample through, thus accumulating more analyte in one place and improving detection [6]. SERS-active particles can also be optically trapped as discussed above and magnetic particles can be trapped using a magnet. The magnets used are often simple magnets or electromagnets but coverage can be uneven with particles sticking to the edges. It is possible to design an electromagnet using similar technology to that used in making lab-on-a-chip devices to create an optical gradient which accumulates the particles at a small spot in the centre increasing sensitivity and reliability [5] (Figure 7.3).

Figure 7.4 shows a simple design for the detection of a DNA sequence diagnostic for Chlamydia [6], one of the most common diseases detected using a routine molecular diagnostic test. DNA from the patient is amplified off chip to select out a specific diagnostic oligonucleotide. A sample is then added to the chip channel pre-filled with capture

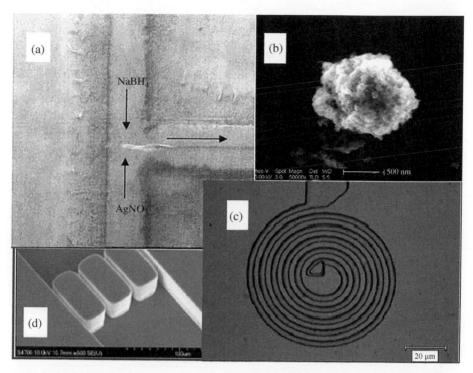

Figure 7.3. A few ways lab-on-a-chip technology can be used to aid on chip SERS detection. (a) The formation of colloid *in situ* to form a thin linear stream giving a fresh reproducible substrate for repeat analysis, (b) a silver-coated magnetic particle which can be trapped in a micro magnet on chip, (c) a magnet designed to give a field gradient which forces the particles into the centre area where the excitation can be focused and (d) pillars added to a channel in a chip to hold beads and allow flow. Source: Reproduced with permission from (a) Keir, R., Igata, E., Arundell, M., et al. (2002). *Anal. Chem.* **74**: 1503 [4]. Copyright American Chemical Society (b) and (c) Quinn, E.J., Hernandez-Santana, A., Hutson, D.M., et al. (2007). *Small* **3**: 1394 [5]. Copyright John Wiley and Sons and (d) Monaghan, P.B., McCarney, K.M., Ricketts, A., et al. (2007). *Anal. Chem.* **79**: 2844 [6]. Copyright American Chemical Society.

beads held against pillars where the specific oligonucleotide is retained. A complimentary oligonucleotide with a SERRS label is then hybridised to the oligonucleotide and the beads thoroughly washed. On heating, the labelled oligonucleotide is released, mixed with colloid and the SERRS signal detected downstream giving quantitative results.

The sensitivity which can be achieved is very high and the assays can be very reproducible. However, the input sample is very small and this can create a problem. There is little point in claiming high sensitivity in terms of the number of molecules detected if, given a 2 ml plasma sample, 1 µl of sample is added to the device. Other techniques may use a larger proportion of the sample to obtain the same sensitivity. In some samples, it is important to ensure homogeneity and again small sample size

(a)   (b)

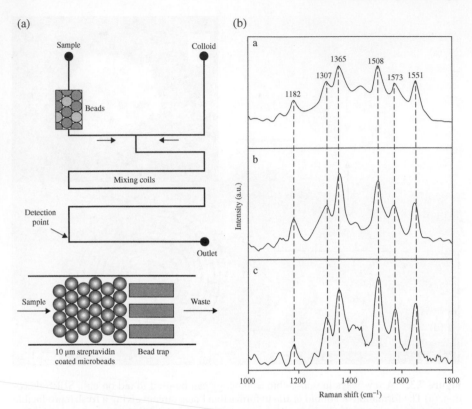

Figure 7.4. (a) A simple chip design for the detection of Chlamydia. Amplified DNA using PCR is captured on the beads, identified by hybridisation with a complimentary oligonucleotide labelled with a SERRS-active label, washed, the labelled nucleotide released by heating and mixed with colloid for detection. (b) SERRS from the labelled oligonucleotide. a: taken off chip, b: on chip with no Chlamydia and thus no capture and c: after capture and washing. Source: Reproduced with permission from Quinn, E.J., Hernandez-Santana, A., Hutson, D.M., et al. (2007). *Small* **3**: 1394 [5]. Copyright John Wiley and Sons.

can be a problem. There is also a barrier to some in the cost and difficulty of setting up a good system and in some cases a less elegant but simpler flow system made with tubes is easier to develop, cheaper and quite satisfactory. Overall, however, once developed, the stability of the chip method and the high sensitivity obtained make this a promising technique. A review describes some more complex devices [7].

7.3 SPATIAL RESOLUTION

Raman spectrometers coupled to microscopes enable samples to be imaged or mapped down to less than the diffraction limit of $1/2\lambda$ using algorithms and scanning methods. For good imaging, the microscope is set up confocally so that only light from the plane at the focus is collected efficiently. Simple imaging and mapping methods have been described in Chapter 2 but much more information can be obtained more quickly including 3D images by using hyperspectral imaging. There are various ways of achieving this but a common method is to present the exciting radiation as a line across the sample and focus the scattering as a line on a 2D detector such as a CCD chip. This enables the Raman scattering to be recorded in the direction perpendicular to the line. The line is then either swept across the sample or the sample stage moved to record a 2D Raman image at regular distances across the sample. Repeating the scan at various depths enables a 3D picture to be obtained.

This works well and there are many good examples. One great strength of Raman scattering is that the sharp nature of the spectra means strongly scattering components of samples can be identified *in situ* in the images with no further processing. A good example of this are lipids in cells which have strong C—H stretches. Added labels including dyes and pigments which can be resonantly enhanced can help identify specific targets. However, the enhancement tends to be less than for other techniques such as fluorescence, so where sensitivity is critical, SERS-active nanoparticles are often used.

Figure 7.5 shows an example of a 3D image identifying SERS-active nanoparticles [8]. Four types of SERS-active nanoparticles with different labels were incubated with Chinese Hamster Ovarian (CHO) cells. The images show a false colour Raman image of part of three cells, where small aggregates of nanoparticles are present (dark dots in red boxes in Figure 7.5a). The expanded image shows the depth profiling. Three of the four types of nanoparticles are shown to be present in different amounts at different depths.

There are other ways that a good Raman spectrum can be obtained from nanoscale samples. For example, if a nanoparticle particle can be isolated and adsorbed on a surface so that no other effective Raman scatterer lies within a few microns of it, Raman scattering can be recorded under a microscope but no detail of the sample geometry can be obtained. However, the sample can be relocated and scanned using TEM, SEM, AFM or STM. Alternatively, the experiment could be done in a combined system. The Raman scattering is measured from only a small part of the area illuminated by the focused laser beam with the result there can be significant background radiation collected and it is very tempting to increase the laser power. This technique works well with SERS-active particles and an example of its use is detailed in Chapter 5. However, higher laser powers can cause sample drying and photodegradation leading to some unusual spectra and some papers dealing with plasmonic substrates report plasmon assisted catalysis where specific reaction products can be made.

Scanning near-field optical microscopy (SNOM) uses radiation which has been confined. In one form of this technique a glass fibre is coated with aluminium or another metal. This is then heated and pulled to narrow the diameter of the fibre and cleaved at the narrowest part to leave an optically clear small aperture. When light is launched down the fibre, it is contained by the metal coating so that the amplitude of the light is compressed within the tube as it narrows. When the light emerges from

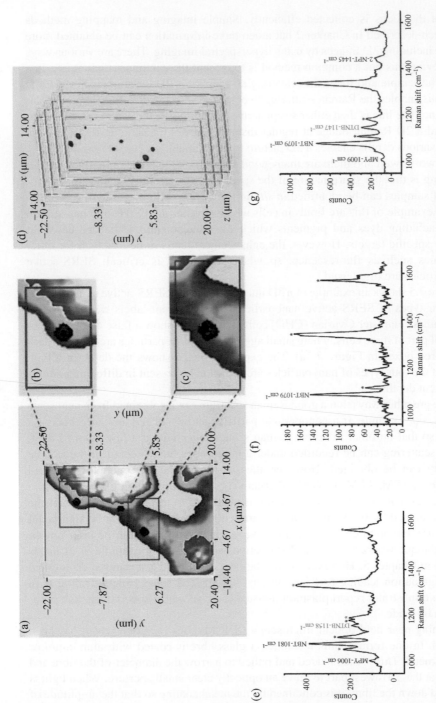

Figure 7.5. An illustration of 3D imaging. (a) Parts of three CHO cells showing small clusters of labelled nanoparticles, (b) and (c) expanded view of the clusters and their positions, (d) slices taken at different depths. (e, f and g) SERS taken at different depths showing peaks from different labels and indicating different distributions of the labelled particles at each depth. Source: Reproduced from McAughtrie, S., Lau, K., Faulds, K., and Graham, D. (2013). *Chem. Sci.* **4**: 3566 [8] with permission from the Royal Society of Chemistry.

the aperture it can be as narrow as 50 nm. If the fibre is placed so that the aperture is almost in contact with the surface, the effective irradiation area is about a 50 nm circle. The way in which this is done in practice is usually to adapt the technology developed for atomic force microscopy (AFM) to position the fibre and then to scan the sample. The Raman scattering is collected from the area containing the spot. There is no need for the collection optics to be focused down to 50 nm since the Raman scattering will only be from the area covered by the spot. The very small excitation volume and the difficulty of collection from close to the tip means that longer accumulation times may be required than would be normal when using a Raman microscope to study a comparable scatterer.

7.4 PULSED AND TUNABLE LASERS

Much of the work described so far can be achieved with continuous wave (CW) lasers which continuously emit light of a specific frequency. However, both tunable and pulsed lasers can have specific advantages as excitation sources in Raman scattering and are widely available. Much of the physics for the development of these systems is not new, but what is new is that many more reliable lasers with different emission frequencies, pulse shapes and repetition rates are available and that tunable systems are simpler and more reliable. Together with improved optics and detectors, this has radically changed the utility of pulsed systems so that usage is increasing significantly. In particular, the wider availability of high-repetition rate low-peak power systems has reduced the problem of photodecomposition and fast detectors have enabled the direct recording of scattering even with picosecond excitation.

The choice of frequency is dependent on use but with fast systems, one fundamental to be considered is the Heisenberg uncertainly principle

$$\Delta E \times \Delta t = \frac{h}{2\pi}$$

When Δt is very short as in a femtosecond system, ΔE is large so that the pulse emitted has a wide frequency range, too wide to obtain sharp Raman bands directly. However, for example, with two photon techniques a femtosecond pulse can be used with a picosecond probe to achieve the wavelength selectivity or the intensity of the Raman spectrum can be recorded against time. In the latter case, a Fourier transform can then applied to obtain the frequency dependence.

One advantage of pulsed lasers is that phase-sensitive detection can be used. This works by comparing the light levels with the pulse on against the level with it off and averaging the difference in value over many pulses. A simple example of an advantage of this would be with Raman scattering in ambient light where the background would be present in both on and off signals eliminating it. However, it is with the nonlinear techniques that the advantages of both efficient background rejection and high sensitivity are used more fully. There are many stable, narrow band, long-life lasers which are available. A common one is a solid-state laser based on neodymium (Nd) ions doped into a yttrium aluminium garnet (YAG) crystal which emits at 1064 nm or, with the frequency doubled, at 532 nm. The beam from the laser can be tuned with an optical paramagnetic oscillator (OPO). The combination is well established and can be built into a modern system reliably. None of this is revolutionary, but improvements throughout mean that modern systems work well and enable users to carry out more ambitious experiments more easily.

In this book so far, Raman scattering has been described in terms of a single photon event in which the Raman scattering efficiency is linearly related to the laser power. However, what happens at higher power densities if no photodecomposition occurs? In this case more than one photon may interact with any one molecule at the same time causing a multiphoton event, the magnitude of which is not linearly related to the laser power. This condition is relatively easily achieved using pulsed lasers. There are a range of techniques which use this effect with different advantages. For example, hyper Raman spectroscopy (HRS) has different selection rules from Raman scattering. CARS also has different selection rules and partly overcomes the problem of taking measurements in fluorescent media. SRS also overcomes some interference problems. A short description of these three techniques is given to illustrate the potential of this type of technology.

In HRS, an intense pulsed beam of radiation is focused onto the sample. One common way of achieving this is to use a 1064 nm Nd YAG laser. If sufficient power is present in the pulse, two photons may interact with the one molecule to create a virtual state at double the frequency of the laser excitation. Raman scattering from this virtual state to an excited vibrational state of the ground state then occurs (Figure 7.6). The scattering depends on hyper polarisability (β) not polarisability (α). In Chapter 1 the use of Hooke's law was discussed. It was pointed out that in a molecule, the shape of the curve which describes the effect of internuclear

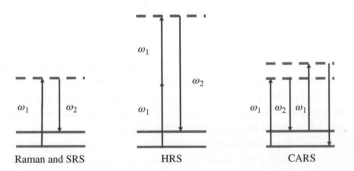

Figure 7.6. The basic processes of Raman scattering, HRS and CARS. In CARS both upward transitions (ω_1) are shown as the same frequency and are normally from the same laser. This is commonly the case but it is not a requirement. The diagram for SRS is the same as for Raman scattering although the scattering intensity will be much greater.

separation on the states of a molecule deviates from the parabola which is used to derive Hooke's law. To accommodate this, second and higher terms are added to correct for this. Similarly, polarisability is usually expressed as a series of terms, the first term being the basic polarisability term. The second term is the hyper polarisability term and it produces different selection rules. Intense scattering is obtained from less-symmetric vibrations and from some vibrations not intense in either Raman scattering or infrared absorption. By comparing the Raman and hyper Raman scattering, more can be understood about a specific analyte. The main disadvantage is that HRS is inefficient with a much lower probability of a scattering event occurring than for normal Raman scattering.

Surface enhanced hyper Raman scattering (SEHRS) is one way of overcoming the inefficiency of the HRS process with very high enhancement factors claimed [9, 10]. Pyridine enhancement has been calculated as about 10^{13} and some dyes have enhancement factors of about 10^{20}! Figure 7.7 shows a SEHRS spectrum for pyrazine, a substance widely used to probe the SERS effect [11]. Pyrazine has a centre of symmetry and consequently, in Raman scattering, only symmetric vibrations will appear in the spectrum. However, in SERS more vibrations appear since the adsorption to the surface

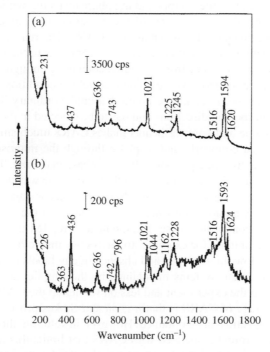

Figure 7.7. SERS (a) and SEHRS (b) of pyrazine on a silver electrode. Source: Reproduced from Li, W.H., Li, X.Y., and Yu, N.T.U. (1999). *Chem. Phys. Lett.* **305**: 303 [11] with permission from Elsevier.

breaks the centre of symmetry. With SEHRS even more bands appear illustrating the different selection rules. When a chromophore is attached to a molecule as a label, very intense spectra can be obtained as the example for a carotene shows [12] (Figure 7.8). In this case the SERRS taken at 532 nm is quite similar to the SEHRS spectrum with some intensity changes and both are shown as high-quality spectra compared to SERS.

The most widely used nonlinear technique to date has been CARS. Reviews on this technique can be found in Refs. [13–15]. A photon creates a virtual state as for ordinary Raman scattering (ω_1 in Figure 7.6). The frequency of the second photon (ω_2) is chosen to stimulate depopulation of the virtual state and return the molecule to an excited vibrational state of the ground state which is instantly heavily populated. A third photon is then used to excite the molecule to a second virtual state. Scattering from this second virtual state returns the molecule to the ground state and it is this radiation which is collected. To obtain a spectrum the frequency of ω_2 is varied to populate each vibration in turn. CARS is usually simplified by using only two sources. In this case the upward photons are from the same laser and are therefore shown as the same frequency ω_1 in Figure 7.6 but it is possible to use a different frequency for the second upward transition.

Since CARS uses at least two different frequencies of radiation, the pulses need to arrive simultaneously at the sample making phase matching without a microscope a problem. Originally the beams were either colinear or at an angle, typically about 7°. This type of arrangement is known as BOXCARS. The reason for using angled beams is that the interacting length between the beams where the pulses combine is shorter making collection easier from a volume containing enough molecules undergoing the CARS process. However, the phase matching condition is quite complex and difficult to calculate. It should be noted that unlike ordinary Raman scattering, CARS is emitted in specific directions and has to be detected in these directions.

Using a microscope and colinear beams makes CARS much simpler. Light from two lasers is directed colinearly and in phase through the microscope optics. The sharp focus of the microscope provides effective phase matching. This development together with modern solid-state lasers and OPO to tune the wavelength of ω_2 makes CARS more accessible and simpler to use. In the simplest arrangement, the detector is usually a large area photo diode which gives an image of the one vibration which satisfies the CARS condition. These developments have led to substantial growth in the use of the technique, for example, to analyse or image tissue samples where the combination of low fluorescence and absence of any labels is of value. It is now possible to buy complete systems maintained by the supplier so that the user can concentrate more on the experiment and less on setting up the CARS system.

The main advantage of CARS is that it is an anti-Stokes process and, as a result, fluorescence-free spectra can be obtained. However, in solution, there is an appreciable electronic background associated with CARS that limits this advantage. CARS has specific selection rules which arise from the third polarisability component and if the spectrum is compared to the spontaneous Raman or resonance Raman spectrum, this can give a more effective assessment of the properties of a molecule.

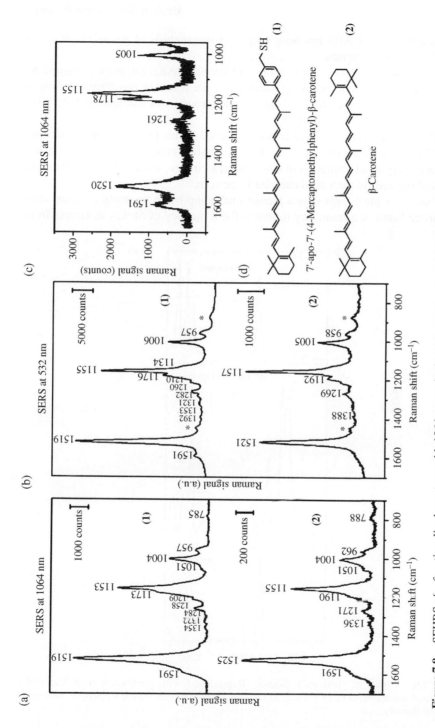

Figure 7.8. SEHRS of a functionalised carotene with 1064 nm excitation (a) compared to SERRS with 532 nm excitation (b). Source: Reproduced from Gühlke, M., Heiner, Z., and Kneipp, J. (2016). *J. Phys. Chem. C* **120**: 20702 [12] with permission from the American Chemical Society).

One example where CARS has been used is in the analysis of gas mixtures in the head of combustion engines.

For many solution-phase applications, additional CARS intensity is gained by using excitation frequencies giving resonant or pre-resonant conditions making it easier to discriminate the signals from the background. An example of CARS for rhodopsin [16] is shown in Figure 7.9.

CARS imaging can be very effective, giving good contrast from specific vibrations. Figure 7.10 shows a section of tissue from a prostate cancer tumour with a blood vessel in it. Very good contrast is obtained from an image of the $2880\,cm^{-1}$ vibration. Individual red cells in the vessel can clearly be seen.

SRS in its simplest form uses a picosecond pump beam to create a virtual state and a probe beam of a frequency to match the frequency of Stokes scattering from

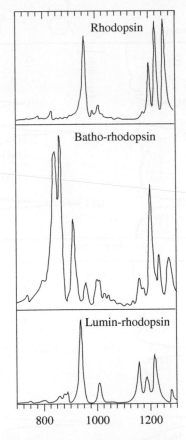

Figure 7.9. CARS of rhodopsin. Source: Reproduced with permission from Yager, F., Ujj, L., and Atkinson, G.H. (1997). *J. Am. Chem. Soc* **119**: 12610 [16] Copyright American Chemical Society.

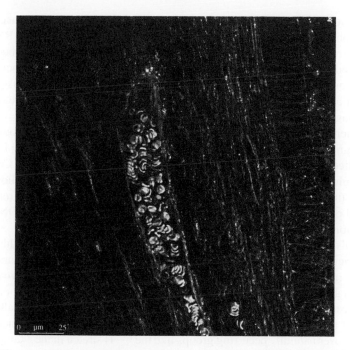

Figure 7.10. A section of prostate cancer tissue imaged by CARS. The tight focus was achieved with a water immersion objective. The pump was at 795 nm and the Stokes probe at 1031 nm. This selects the 2880 cm⁻¹ aliphatic C—H stretch. The individual red cells in the blood vessel can be clearly identified with no staining. Source: Unpublished work reproduced with permission from Jamieson, L., Faulds, K., and Graham, D. Centre for Molecular Nanometrology, Strathclyde University, Glasgow. Sample supplied by Leung H. and Salji M, Beatson Institute for Cancer Research, Glasgow.

the virtual state to a specific vibrational level [17–20]. As with CARS, the Stokes emission is stimulated thus depopulating the virtual state very quickly and increasing the scattering efficiency for the chosen vibration by orders of magnitude. Like CARS and HRS, SRS is a nonlinear process, so sufficient photons from both the pump and probe beams must be present on the molecule at the same time and again, this can be achieved effectively by tightly focusing them through a microscope. To detect the signal, the intensity of the pump and probe beams is compared using phase-sensitive detection and as before the detector is usually a large-area photo diode. Either loss of intensity to the pump beam or intensity gain to the probe beam (SRG) is measured and the sensitivity is claimed as nearly shot noise limited. A diagram of the type shown in Figure 7.6 for other techniques looks the same for SRS as for normal Raman scattering. However, the downward stimulated scattering process is much more efficient.

There are several advantages of this technique. The process is much more efficient than normal Raman scattering (usually called spontaneous Raman scattering in this context) but has the same selection rules. Signal intensity is linear with concentration making concentration dependence easier to study and compared to CARS there is less background. For mapping and imaging, sufficient photons from both beams are required to be present to achieve SRS, so tight focusing with the microscope is advantageous. This means only a very small volume will give SRS, thus providing good area and depth discrimination. Combined with the large-area photodiode this makes imaging very effective.

The increased efficiency of the process compared to spontaneous Raman scattering makes it effective for label-free detection. Proteins, lipids and DNA have been studied and cells imaged with good contrast between components. Cell uptake of molecules, which have good Raman cross-sections, have been followed. For example, the dominant C—H stretch from DNA is slightly different from that in lipids and proteins and this has been used to observe the movement of DNA during cell division (anaphase and metaphase stages).

With this high sensitivity, small probes and tags can be identified readily in complex matrices like cells. For example, the —C≡C— bond in alkynes occurs at about $2000\,cm^{-1}$ in a region of the spectrum where molecules like lipids and proteins have no appreciable spectra [21]. Therefore, adding a small alkyne tag to a biomolecule can make it possible to track by SRS. The high cross-section of groups like alkynes and the lack of interference from other bands makes this easy to detect and increases the sensitivity over naturally occurring groups like the C—H stretch in the example above.

So far, these techniques have been described for the collection of scattering from one vibration at a time. However, for CARS and SRS it is possible to obtain complete spectra in one measurement. Using a femtosecond pulse will produce a probe with a narrow time range but a broad frequency range. If a narrow frequency picosecond pulse is then used, all vibrations can scatter simultaneously since the virtual states will span a range of energies due to the broad pump. A spectrometer is then used to separate the different frequencies. Figure 5 of Ref. [19] explains the difference between the two approaches well.

The reader should be aware that each of the three processes described above will occur at the same time but separate detection is possible because the frequency of the scattered light is different. A more subtle point is that other processes also occur. Two-photon absorption and two-photon emission will also occur. Thus, it can be advantageous to use lower frequency beams, otherwise these two-photon processes will occur with allowed UV bands and can create either efficient absorption or pulsed, and therefore detectable, fluorescence.

7.5   TIP-ENHANCED RAMAN SCATTERING AND SNOM

TERS combines the use of SERS enhancement and scanning using an AFM or a STM to give a map of a surface [22, 23]. The cantilever tip of the AFM/STM is coated with silver or gold so that the tip is covered in a rough metal SERS-active

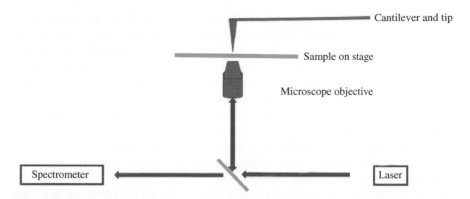

Cantilever and tip

Sample on stage

Microscope objective

Spectrometer ←——— ←——— Laser

Figure 7.11. A simple diagram of a TERS setup. The cantilever carries the SERS-active tip and the scattering is collected from the inverted microscope objective. The stage is moved in a controlled fashion and spectra recorded at preordered intervals.

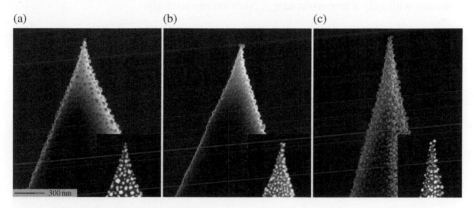

(a)   (b)   (c)

300 nm

Figure 7.12. Images of a tip coated with silver showing tips with one or two silver particles at the apex of each. Source: Taken from Deckert, V., Deckert-Gaudig, T., Diegel, M., et al. (2011). *Faraday Discuss.* **177**: 9 [24] with permission from the Royal Society of Chemistry.

coat. If the tip either contacts or comes close to the surface and the area is illuminated by a laser, strong SERS/SERRS can be obtained. Since this means an enhancement of 10^6 or more, effectively only light from the point where the tip is near the surface is collected. A simple diagram of one arrangement is shown in Figure 7.11. To obtain a map the piezo electric stage will usually be moved rather than the tip. Epi illumination is also used. The AFM image is obtained by scanning across the sample whilst recording the Raman scattering to create a matching Raman map. The resolution is dependent on the parameters set for the AFM and the quality of the tip. The latter can be a limiting factor. Usually in a good system the resolution is about 10–20 nm. Figure 7.12 shows EMs of well-prepared tips with either one or two silver particles at the tip.

Although 10–20 nm resolution is usually a good result, with care and attention and high-quality apparatus, much higher resolution can be obtained. Often, the sample is adsorbed onto a metal surface of silver or gold although high resolution can also be obtained from an insulating substrate. The plasmon formed as the tip approaches the surface is between the particle at the apex of the tip and the metal substrate, confining it to a very small volume. Single-molecule measurements of polypeptides, DNA and other biomolecules have been obtained and bases in RNA identified giving the possibility of sequencing. In 2013, Zhang et al. [25] used a STM in high vacuum and at low temperature to not only obtain the spectrum of a single molecule but to observe differences in the spectrum from different parts of the molecule. The resolution obtained is less than a nanometre (Figure 7.13).

These studies have implications for SERS theory. Given the curvature of the tip, only a few silver atoms are close to the molecule analysed and the field gradient will be generated from only those. The whole process takes place in a much smaller space than most SERS and shows that suitable hot spots can be generated in very small volumes with only a few silver atoms contributing strongly.

7.6 SINGLE-MOLECULE DETECTION

TERS is not the only way that single-molecule detection can be achieved. Kneipp et al. [26] and Nie and Emory [27], early in SERS development, demonstrated that SERS can be observed at the single-molecule level. Since then, the attraction of studying the properties of a single molecule has led to many studies and some good results [28]. However, unlike TERS where the molecule is observed directly, with most methods it is necessary to use the data obtained to show that only one molecule is being studied and this is not always done. For example, if a concentration study is carried out, one problem well known in forensic science is that the molecule can stick strongly to glass. At these very low concentrations this leads to an equilibrium, between the walls of the vessel and the substrate, changing the concentration significantly and leaving a difficult-to-remove residue on the vessels used. There are several studies in the literature claiming detection limits to the single-molecule level where there is either no concentration data reported or concentration data with little intensity change as the concentration is reduced. These results are likely to be due to residual analyte.

A common way of attempting single-molecule measurements with SERS is to adsorb the analyte from a low-concentration solution onto an isolated silver nanoparticle dimer and then to detect the scattering from the dimer and surrounding area using a microscope. However, the molecule giving SERS can adsorb on any part of the dimer. There is sometimes little evidence that only one molecule will adsorb and scatter from the hot spot or that some scattering does not arise from many molecules on other parts of the surface. Various methods are used to prove that signals from this

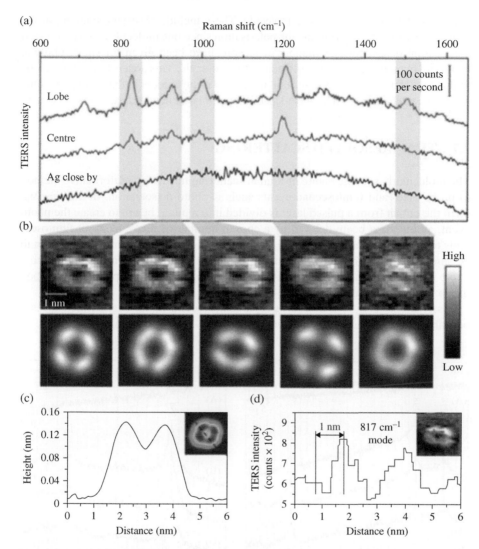

Figure 7.13. TERS mapping of a single meso-tetrakis(3,5-di-tertiarybutylphenyl)-porphyrin (H2TBPP) molecule deposited on silver showing TERS obtained from the lobe (red) and centre (blue) of the molecule (a and b), (c) Gives the height profile of a line trace across the molecule and (d) TERS map of the 817 cm^{-1} intensity for the line shown in (c). Source: Reprinted with permission from Zhang, R., Zhang, Y., Dong, Z.C., et al. (2013). *Nature* **498**: 82 [25]. Copyright Springer Nature.

type of system are from a single molecule. These include observing blinking of the signal with time and determining the polarisation since one molecule can have defined polarisation dependence. A number of reasons have been given for signal blinking including photobleaching and diffusion of the molecule. This area is beyond the scope of this book but this section may be of interest to some readers in that it highlights the need for extreme care and hard evidence if low limits of detection are to be reported.

7.7 TIME-RESOLVED SCATTERING

The molecularly specific nature of Raman scattering makes it an informative method to probe pico- and femtosecond events such as photo-dissociation. In the simplest form the output from a pulsed laser is divided into a pump beam, to create the photo event, and a probe beam delayed by a chosen number of pico-, nano- or femtoseconds to create the Raman scattering. Changing the delay time allows the reaction to be studied as it progresses. The photo-dissociation of carbon monoxide bound to a heme is a good example (Figure 7.14). Good evidence about the pathway of carbon monoxide desorption in the enzyme is obtained.

Faster events can be followed using a femtosecond pump beam which gives a broadband pulse in terms of frequency. The pulse is shorter than the time scale of

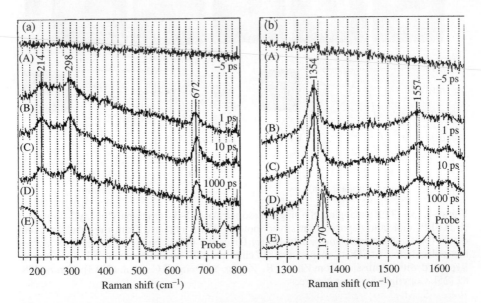

Figure 7.14. The photo-dissociation of CO from the heme centre in myoglobin. The band at 1370 cm⁻¹ in spectrum (E) is the oxidation state marker. Dissociation of the CO from the heme causes reduction and other changes. Source: Reproduced from Sato, A., Sasakura, Y., Sugiyama, S., et al. (2002). *J. Biol. Chem.* **277**: 32650 [29] with permission the American Physical Society).

vibrational motion (low picosecond range). This results in the dephasing of the output to give an oscillating signal which decays with time. A Fourier transform then provides the Raman spectrum. The technology is beyond the scope of this book but the information which can be obtained concerning short-lived species is unique. An example of resonance Raman spectra taken by this method to follow the formation of the singlet state in β carotene is given to illustrate the potential in Figure 7.15. The reader is referred to a review on this subject [31].

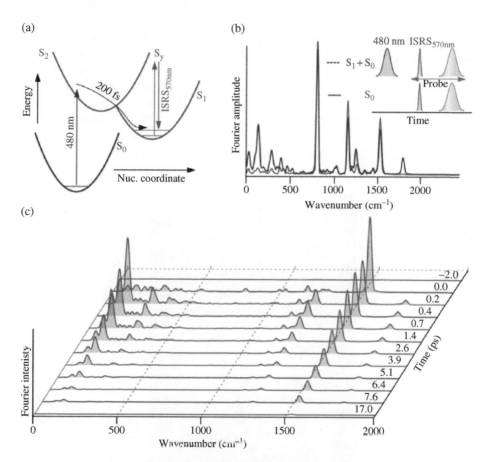

Figure 7.15. (a) Time-resolved Raman scattering obtained with a femtosecond pulse using vibrational coherence for β-carotene in toluene. The femtosecond pulse excites the molecule to state S_1 which then decays into state S_2 in 200 fs. (b) The system is then interrogated by a picosecond probe pulse. Depending on the timing of the pulses either the ground-state Raman scattering or a mixture of the ground- and excited-state Raman scattering can be obtained. (c) This allows by subtraction Raman scattering from the excited state to be obtained. Source: Reproduced from Liebel, M., Schnedermann, C., Wende, T., and Kukura, P. (2015). *J. Phys. Chem. A* **119**: 9506 [30] with permission from the American Chemical Society.

(a)

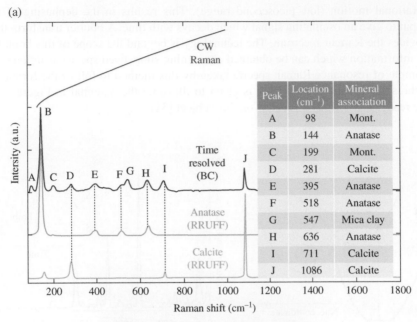

Peak	Location (cm^{-1})	Mineral association
A	98	Mont.
B	144	Anatase
C	199	Mont.
D	281	Calcite
E	395	Anatase
F	518	Anatase
G	547	Mica clay
H	636	Anatase
I	711	Calcite
J	1086	Calcite

(b)

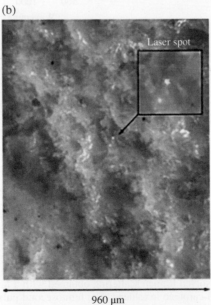

Figure 7.16. Spectrum from a mineral taken with a fast detector and with fluorescence rejection applied. The advantage over a normal Raman spectrum is clear. Source: Reproduced from Blacksberg, J., Alerstam, E., Maruyama, Y., et al. (2016). *Appl. Optics* **55**: 739 [32] with permission from Applied Optics.

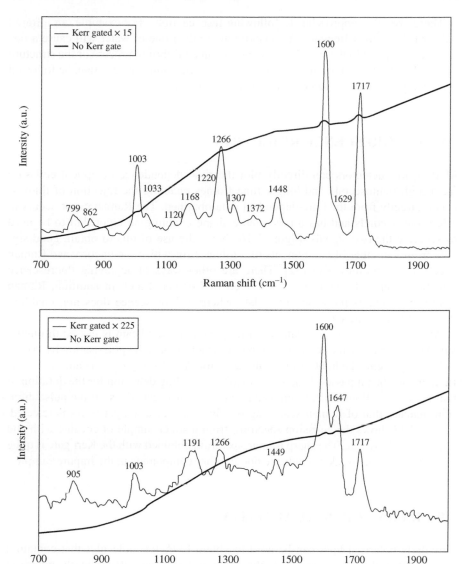

Figure 7.17. Kerr gated (light line) and normal Raman scattering (heavy line) for fluorescent street samples of cocaine. The samples contain cutting agents and cocaine in different concentrations but the cocaine peaks at 1717 100 1266 and about 1003 cm^{-1} are clear. Source: Reproduced from Littleford, R.E., Matousek, P., Towrie, M., et al. (2004). *Analyst* **129**: 505 [33] with permission from the Royal Society of Chemistry.

There are other approaches to following fast reactions. For example, an infrared pulse of the correct frequency to adsorb into an overtone of water will cause a fast temperature jump in the sample, thus altering an equilibrium process or the structure of a molecule. The return of the system to the equilibrium state can then be followed by subsequent pulses timed to fit the systems being studied.

7.8   FLUORESCENCE REJECTION

Modern fast detectors can directly plot the time dependence of optical emission from high femtosecond and longer time frames enabling the rejection of fluorescence directly from the time-dependent data obtained. The Raman event occurs in a few picoseconds and has a predictable shape enabling an algorithm to be used to separate the two signals. Figure 7.16 shows the use of this to obtain good signals from a number of components of a mineral for which the uncorrected Raman spectrum is uninformative [32]. There are other ways of achieving fluorescence rejection, some of which were described in Chapter 2 and, in addition, Raman scattering contains phase-sensitive data whereas fluorescence does not, enabling the two to be separated.

Alternatively, a Kerr gate can be set up [33]. A pulsed beam is passed through a suitable medium such as carbon disulphide which rotates the polarisation by 90°. A polariser prevents the light reaching the detector. A following pulse beam rotates the light again when it passes through the medium enabling detection for the duration of that pulse only. Raman scattering occurs in a few picoseconds, so if the pulse duration matches that of Raman scattering the fluorescence at longer times is rejected. Figure 7.17 shows the emission spectrum from a street sample of cocaine with and without the Kerr gate. The sharp Raman spectrum obtained with the Kerr gate is quite clear. Some of the weaker bands are due to other compounds in the impure sample.

7.9   RAMAN OPTICAL ACTIVITY

Circularly polarised light can be used to obtain ROA for molecules that contain a chiral centre. What is measured is the difference between left circularly polarised and right circularly polarised scattering. The difference in intensity is ROA [34, 35]. Two methods are usually used to achieve this. In incident ROA (ICP ROA), alternating left circularly polarised and right circularly polarised pulses of light excite the chiral analyte (Figure 7.18). In scattered ROA (SCP ROA) which is more often used now, the incident light which is usually linearly polarised excites the analyte and the chiral scattering component from it is detected. The selection rules require consideration of both the electric and magnetic dipole operators.

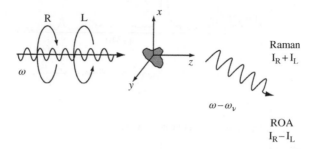

Figure 7.18. Schematic diagram of the basic concept of ICP ROA. What is measured is a small difference in the intensity of Raman scattering from chiral molecules in right (I_R) and left (I_L) circularly polarised light. Incident photons have angular frequency ω and Raman scattered photons have angular frequency $\omega - \omega_v$. For achiral molecules the ROA intensity is zero. Source: Reproduced with permission from, Professor Laurence Barron, University of Glasgow, Scotland.

ROA has been used to study such topics as the absolute chirality of drugs and the chirality in different environments of amino acids, peptides, proteins, DNA and RNA bases [36, 37]. For example, it can selectively identify particular features within a protein such as the degree and type of folding [35] or bulges and mismatches in RNA [38]. Initially the signals were weak and measurement took time but as with other techniques this has improved with equipment advances. An example of the use to determine the chirality of polypeptides under different conditions is shown in Figure 7.19.

7.10 UV EXCITATION

There are some real advantages to using UV excitation. The fourth power law for scattering makes UV Raman scattering much more sensitive than visible Raman scattering and the shorter wavelength reduces the diffraction limit so that greater spatial resolution is possible. Many materials adsorb in the UV and will be resonantly enhanced if the correct wavelength is chosen thus further increasing sensitivity and selectivity compared to visible and near-IR excitation. Fluorescence interference is less of a problem since the emission, if any, is at much lower energies.

There is a drawback, however, in that the high-energy radiation and the presence of many chromophores make photodecomposition an even more serious issue than with visible excitation. To mitigate this problem liquid samples are often presented in a flow cell so that each individual molecule is only interrogated once or if

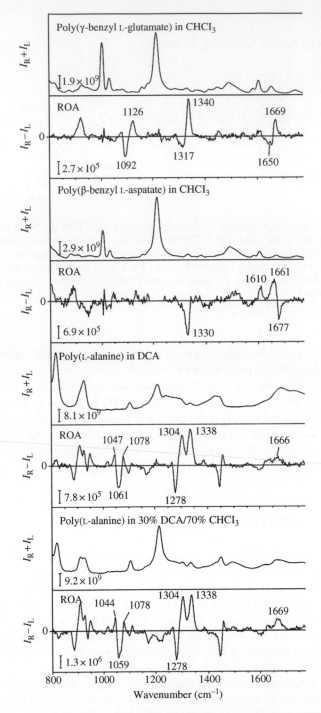

Figure 7.19. Shows the Raman and ROA spectra from polypeptides in α helix configuration. The ROA spectra can be either positive or negative depending on whether the rcp or lcp light is more strongly rotated. The helices are right handed except for poly(β-benzyl L-aspartate) which is a left-hand helix. Source: Reproduced with permission from McColl, H., Blanch, E.W., Hecht, L., and Barron, L.D. (2004). *J. Am. Chem. Soc.* **126**: 8181 [37]. Copyright (2004) the American Chemical Society.

the sample is recycled it has time to dissipate energy between excitation events (see Chapter 2). Solid samples are often presented as disks which can be spun and interrogated on one side so that the sample is effectively a circular track. Further, by using part of the advantage gained from the increased efficiency, lower excitation powers can be used.

As with other Raman methodologies, UV Raman is continuously developing due to improved lasers, optics and detectors. The sensitivity advantage can be used, for example, to improve detection limits for samples such as the explosive PETN which, because of its low vapour pressure, can be difficult to identify in small quantities. Standoff detection systems [39] have been described and targets such as explosives and drugs detected successfully. The presence of many chromophores can be a significant advantage in studying complex systems selectively. Figure 7.20 shows the effect of changing the excitation wavelength for tyrosines indicating resonance enhancement with excitation in the 230 nm region.

For proteins, the resonance effect can be used to select specific groups which are sensitive to their environment [38]. For example, excitation at 229 nm will pick out tryptophan and tyrosines and excitation at 206 nm will pick out vibrations from amide groups because there is an allowed transition from the amide bond at about 190 nm. These bands are structure sensitive. The amide bands, in particular, have been well studied making it possible to differentiate between structures like α helices and β sheets and to study protein folding. Oladepo and coworkers [41] developed this so that secondary structure of proteins could be obtained quantitatively directly from the Raman spectra. Excitation at 244 nm picks out vibrations from DNA bases, as well as aromatic side groups, from amino acids and this has been used in more applied studies, for example, to monitor changes in RNA and DNA in culture medium used with mammalian cells [42].

As an example of the rich information possible, Figure 7.21 shows the Raman spectrum from myoglobin. Excitation at 413 nm is in resonance with the Soret band of the heme system and picks out bands from it. These give information on the oxidation state and spin state of the heme and some information on the structure due to strain in vinyl groups attached to it. Excitation at 229 nm picks out tryptophan and tyrosine groups which are sensitive to changes in their environment and the informative amide bands can be picked out using 206.5 nm excitation.

This is only a small selection of what has been achieved and much more is possible with modern equipment. For example, some early experiments report eight-hour accumulation times with low power to prevent photodecomposition whereas most modern studies report accumulation times of a minute or less. Further, there are other accessible wavelengths where resonance enhancement will occur such as 280 nm and, with nitro groups, 300 nm.

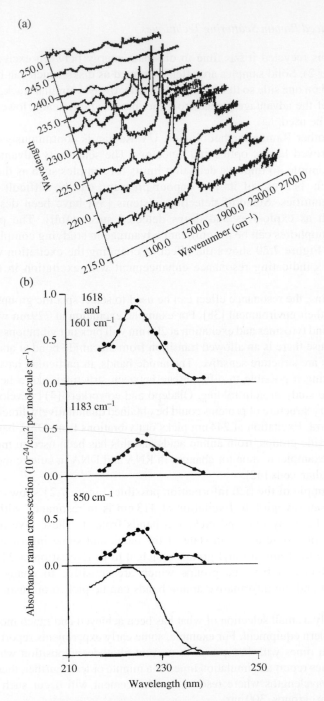

Figure 7.20. UV-resonance Raman scattering for tyrosines. (a) Actual spectra with different excitation wavelengths. (b) Intensity dependence on excitation wavelength of individual peaks. Source: Reproduced with permission from Ludwig, M. and Asher, S.A. (1988). *J. Am. Chem. Soc.* **110**: 1005 [40]. Copyright (1998) the American Chemical Society.

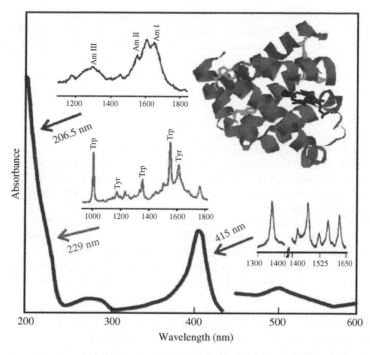

Figure 7.21. Resonance Raman spectra from myoglobin at three different frequencies showing the selectivity of the resonance effect. Source: Reproduced with permission from Oladepo, S.A., Xiong, K., Hong, Z., et al. (2012). *Chem. Rev.* **112**: 2604 [41]. Copyright (2012) the American Chemical Society.

7.11 SUMMARY

Optical developments have resulted in a greater use of Raman scattering. The development of very simple-to-use systems such as the portable systems described early in this chapter are a prime example. In addition, coupling Raman detectors to other instruments, standoff detection either to detect an analyte in a lab-on-a-chip device or at many meters and the use of Raman scattering in hostile environments have become much easier and more effective. The more advanced techniques described later tend to be less available but they offer real advantages and are becoming simpler to use and less expensive. They offer new ways in which Raman scattering can expand and become more powerful in the future.

REFERENCES

1. Cardell, C. and Guerra, I. (2016). *Trends Analyt. Chem.* **77**: 156.
2. Yuan, Y., Lin, Y., Gu, B. et al. (2017). *Coord. Chem. Rev.* **339**: 138.
3. Lehmuskero, A., Johansson, P., Rubinsztein-Dunlop, H. et al. (2015). *ACS Nano* **9**: 3453.

4. Keir, R., Igata, E., Arundell, M. et al. (2002). *Anal. Chem.* **74**: 1503.
5. Quinn, E.J., Hernandez-Santana, A., Hutson, D.M. et al. (2007). *Small* **3**: 1394.
6. Monaghan, P.B., McCarney, K.M., Ricketts, A. et al. (2007). *Anal. Chem.* **79**: 2844.
7. Jahn, I.J., Žukovskaja, O., Zheng, X.-S. et al. (2017). *Analyst* **142**: 1022.
8. McAughtrie, S., Lau, K., Faulds, K., and Graham, D. (2013). *Chem. Sci.* **4**: 3566.
9. Madzharova, F., Heiner, Z., and Kneipp, J. (2017). *Chem. Soc. Rev.* **46**: 3980.
10. Butet, J. and Martin, O.J.F. (2015). *J. Phys. Chem. C* **119**: 15547.
11. Li, W.H., Li, X.Y., and Yu, N.T.U. (1999). *Chem. Phys. Lett.* **305**: 303.
12. Gühlke, M., Heiner, Z., and Kneipp, J. (2016). *J. Phys. Chem. C* **120**: 20702.
13. Schie, I.W., Krafft, C., and Popp, J. (2015). *Analyst* **140**: 3897.
14. Day, J.P.R., Domke, K.F., Rago, G. et al. (2011). *J. Phys. Chem. B* **115**: 7713.
15. Zumbusch, A., Holtom, G.R., and Xie, X.S. (1999). *Phys. Rev. Lett.* **82**: 4142.
16. Yager, F., Ujj, L., and Atkinson, G.H. (1997). *J. Am. Chem. Soc.* **119**: 12610.
17. Freudiger, C.W., Min, W., Saar, B.G. et al. (2008). *Science* **322**: 1857.
18. Fu, D., Lu, F.-K., Zhang, X. et al. (2012). *J. Am. Chem. Soc.* **134**: 3623.
19. Prince, R.C., Frontiera, R.R., and Potma, E.O. (2017). *Chem. Rev.* **117**: 5070.
20. Wei, L., Shen, Y., Xu, F. et al. (2015). *ACS Chem. Biol.* **10**: 901.
21. Zrimsek, A.B., Chiang, N., Mattei, M. et al. (2016). *Acc. Chem. Res.* **49**: 2023.
22. Richard-Lacroix, M., Zhang, Y., Dong, Z., and Deckert, V. (2017). *Chem. Soc. Rev.* **46**: 3922.
23. Zaleski, S., Wilson, A.J., Mattei, M. et al. (2016). *Acc. Chem. Res.* **49**: 2023.
24. Deckert, V., Deckert-Gaudig, T., Diegel, M. et al. (2015). *Faraday Discuss.* **177**: 9.
25. Zhang, R., Zhang, Y., Dong, Z.C. et al. (2013). *Nature* **498**: 82.
26. Kneipp, K.K., Wang, Y., Kneipp, H. et al. (1997). *Phys. Rev. Lett.* **78**: 1667.
27. Nie, S. and Emory, S.R. (1997). *Science* **275**: 1102.
28. Pozzi, E.A., Goubert, G., Chiang, N. et al. (2017). *Chem. Rev.* **117**: 4961.
29. Sato, A., Sasakura, Y., Sugiyama, S. et al. (2002). *J. Biol. Chem.* **277**: 32650.
30. Liebel, M., Schnedermann, C., Wende, T., and Kukura, P. (2015). *J. Phys. Chem. A* **119**: 9506.
31. Sahoo, S.K., Umpathy, S., and Parker, A.W. (2011). *Appl. Spectrosc.* **65**: 1087.
32. Blacksberg, J., Alerstam, E., Maruyama, Y. et al. (2016). *Appl. Optics* **55**: 739.
33. Littleford, R.E., Matousek, P., Towrie, M. et al. (2004). *Analyst* **129**: 505.
34. He, Y., Wang, B., and Dukor, R.K. (2011). *Appl. Spectrosc.* **65**: 699.
35. Barron, L.D. (2015). *Biomed. Spectrosc. Imaging* **4**: 223.
36. Ostovarpour, S. and Blanch, E.W. (2012). *Appl. Spectrosc.* **66**: 289.
37. McColl, I.H., Blanch, E.W., Hecht, L., and Barron, L.D. (2004). *J. Am. Chem. Soc.* **126**: 8181.
38. Hobro, A.J., Rouhi, M., Blanch, E.W., and Conn, G.L. (2007). *Nucleic Acids Res.* **35**: 1169.
39. Gares, K.L., Hufziger, K.T., Bykov, S.V., and Asher, S.A. (2017). *Appl. Spectrosc.* **2017**: 7173.
40. Ludwig, M. and Asher, S.A. (1988). *J. Am. Chem. Soc.* **110**: 1005.
41. Oladepo, S.A., Xiong, K., Hong, Z. et al. (2012). *Chem. Rev.* **112**: 2604.
42. Ashton, L., Hogwood, C.E.M., Tait, A.S. et al. (2015). *J. Chem. Technol. Biotechnol.* **90**: 237.

Appendix A

Table of Inorganic Band Positions

Table A.1. List of band positions in cm⁻¹ observed in some common inorganic compounds. Bold type indicates the strongest bands.

Ammonium	Carbamate	**1039**							
Diamond	Carbon	**1331**							
Ammonium	Carbonate	**1044**							
Calcium	Carbonate	**1087**	713	282					
Lead (II)	Carbonate	1479	1365	**1055**					
Potassium	Carbonate	**3098**	1062						
Strontium	Carbonate	**1072**							
Potassium	Carbonate (99.995%)	**1062**	687						
Potassium	Carbonate (99.995%), rotator	**1061**	686						
Sodium	Carbonate (anhydrous)	**1069**							
Sodium	Carbonate (anhydrous), rotator	**1080**	701						
Sodium	Carbonate AR	1607	**1080**	1062					
Sodium	Carbonate monohydrate	**1070**							
Potassium	Carbonate (99.995%), rotator	**1061**							
Potassium	Carbonate, Aldrich, 99%	**3098**	1062						
Sodium	Chloramine-T, sodium salt	3069	2921	1600	1379	1213	**1132**	930	800
Sodium	Dichloroisocyanurate	1733	1051	707	577	**365**	230		
Potassium	Dichromate	**909**	**571**	387	235				
Sodium	Dichromate (2H₂O)	**908**	371	236					
Potassium	Dichromate, rotator	**909**	570	374	235				
Potassium	Dichromate, rotator	**909**	570	374	235				

(Continued)

Modern Raman Spectroscopy: A Practical Approach, Second Edition. Ewen Smith and Geoffrey Dent.
© 2019 John Wiley & Sons Ltd. Published 2019 by John Wiley & Sons Ltd.

Table A.1.   (*Continued*)

Ammonium	Dihydrogen orthophosphate	**925**				
Ammonium	Dihydrogen orthophosphate	**923**				
Potassium	Dihydrogen orthophosphate	**915**				
Titanium	Dioxide (anatase)	**639**	516	398		
Titanium	Dioxide (rutile)	610	**448**	237		
Sodium	Dithionite	1033	364	**258**		
Sodium	Dithionite	1033	364	**258**		
Ammonium	Ferrous sulphate ($6H_2O$), rotator	**982**	613	453		
Sodium	Hexametaphosphate	**1162**				
Ammonium	Hydrogen carbonate	**1045**				
Caesium	Hydrogen carbonate	**1012**	671	634		
Potassium	Hydrogen carbonate	1281	**1030**	677	636	193
Sodium	Hydrogen carbonate	1269	**1046**	686		
Di-ammonium	Hydrogen orthophosphate	**948**				
Di-ammonium	Hydrogen orthophosphate	**948**				
Di-sodium	Hydrogen orthophosphate	1131	1065	**934**	560	
Di-potassium	Hydrogen orthophosphate (trihydrate)	1048	**950**	879	556	
Potassium	Hydrogen sulphate	1101	**1027**	855	581	412	327
Sodium	Hydrogen sulphate	**1065**	1004	868	601	
Sodium	Hydrogen sulphate (monohydrate)	**1039**	857	603	412	
Calcium	Hydroxide	1086	**358**			
Sodium	Hydroxide	**205**				
Lithium	Hydroxide (monohydrate)	1090	839	517	397	**213**
Ammonium	Hydroxy chloride	1495	**1001**			
Potassium	Iodate	**754**				
Sodium	Metabisulphite	**1064**	660	433	275	
Barium	Nitrate	**1048**	733			
Bismuth	Nitrate	**1037**				
Lanthanum	Nitrate	**1046**	739			
Lithium	Nitrate	**1384**	**1070**	735	237	
Potassium	Nitrate	**1051**	716			
Silver	Nitrate	**1046**				
Sodium	Nitrate	1386	**1068**	725	193	
Magnesium	Nitrate ($6H_2O$)	**1060**				
Iron(III)	Nitrate ($9H_2O$)	**1046**				
Potassium	Nitrite	1322	**806**			
Silver	Nitrite	**1045**				
Sodium	Nitrite	**1327**	828			

Table A.1. (*Continued*)

Sodium	Nitrite	**1327**	828						
Silver	Nitrite, rotator	**1045**	847						
Sodium	Nitroprusside (2H$_2$O)	2174	1946	1068	656	**471**			
Tri-potassium	Orthophosphate	**1062**	940						
Tri-sodium	Orthophosphate	**941**	415						
Tri-sodium	Orthophosphate	1005	**940**	548	417				
Tri-potassium	Orthophosphate	**1062**	972	857	549				
Tri-sodium	Orthophosphate (12H$_2$O)	**939**	407						
Tri-sodium	Orthophosphate (12H$_2$O)	**940**	550	413					
Tri-potassium	Orthophosphate (H$_2$O)	**1061**	939						
Tri-potassium	Orthophosphate (H$_2$O), rotator	**1061**	940						
Cupric	Oxide	**296**							
Zinc	Oxide	**438**							
Cupric	Oxide, rotator	**297**							
Zinc	Oxide, rotator	**439**							
Magnesium	Perchlorate	**964**	643	456					
Ammonium	Persulphate	**1072**	805						
Potassium	Persulphate	1292	**1082**	814					
Sodium	Persulphate	1294	**1089**	853					
Sodium	Phosphate	**938**							
Calcium	Silicate	**983**	578	373					
Lithium	Silicate	**601**							
Zirconium	Silicate	3019	**2821**	2662	1004	438	355	197	
Lithium	Sllicate	**589**							
Calcium	Silicate hydrous, commercial	**983**	578	372					
Magnesium	Rotator	677	**195**						
Magnesium	Silicate hydrous (talc)	676	**194**						
Magnesium	Silicate hydrous (talc), rotator	677	362	**195**					
Aluminium	Silicate hydroxide (kaolin)	**466**							
Aluminium	Silicate hydroxide (kaolin), rotator	912	791	752	705	**473**	430	338	276
Ammonium	Sulphate	**975**							
Barium	Sulphate	**988**	454						
Barium	Sulphate	**988**	462						
Calcium	Sulphate	1129	**1017**	676	628	609	500		
Magnesium	Sulphate	**984**							
Potassium	Sulphate	1146	**984**	618	453				
Silver	Sulphate	**969**							
Sodium	Sulphate (anhydrous)	**993**							
Calcium	Sulphate (dihydrate)	1135	**1009**	669	629	491	415		

(*Continued*)

Table A.1. *(Continued)*

Zinc	Sulphate (heptahydrate)	**985**							
Barium	Sulphate, Raman microscope	**986**	458						
Barium	Sulphate, rotator	**988**	462						
Barium	Sulphate, static	**988**	462						
Sodium	Sulphite	**987**	950	639	497				
Potassium	Sulphite	**988**	627	482					
Magnesium	Thiosulphate (hexahydrate)	1165	1000	659	**439**				
Sulphur	—		471	**216**	151				
Barium	Thiosulphate	1004	687	**466**	354				
Potassium	Thiosulphate (hydrate)	1164	1000	667	**446**	347			
Sodium	Thiosulphate (pentahydrate)	1018	**434**						
Potassium	Titanium oxalate	1751	1386	1252	850	**530**	425	352	300
Potassium	Titanium oxalate ($2H_2O$)	1751	1384	1253	851	**526**	417	353	299

Index

Modern Raman Spectroscopy: A Practical Approach, Second Edition. Ewen Smith and Geoffrey Dent.
© 2019 John Wiley & Sons Ltd. Published 2019 by John Wiley & Sons Ltd.

Index page.

Printed and bound by CPI Group (UK) Ltd, Croydon, CR0 4YY

27/10/2024

14580299-0001